Praise for *Hurricane Lizards and Plastic Squid*

"Thor Hanson's clear-eyed science writing meets its best topic yet in this book. While governments and publics joust over climate change, biologists studying all the ways wild animals are already responding to it are five steps ahead of the game. Hanson takes his readers on a tour of this cutting edge in our rapidly-changing world. Yes, there are looming extinctions. But before you wring your hands in despair, read this book. As it always has, life finds a way."

> —Dan Flores, *New York Times*–bestselling author of
> *Coyote America*

"Hanson writes a hopeful and compelling story exploring various climate adaptations in the animal and plant worlds with a rare combination of engrossing clarity and robust interrogation. He encourages us to lift our own voices and actually assert change. Each enormously engaging essay proves what I've known for some time: Thor Hanson is a marvel whose enthusiasm for this planet is utterly contagious."

> — Aimee Nezhukumatathil, *New York Times*–bestselling
> author of *World of Wonders*

"One of our finest writers of literary natural history takes on the most crucial topic of our times—how will life itself respond to a warming world?—and brings back answers both utterly beguiling and strangely reassuring. This is arguably the most significant discussion of the biology of global warming I know, brought to us in the intelligent, wise, and beautiful prose we've come to rely upon Thor Hanson to deliver. If you read only one book on climate change this year, let it be this one."

> — Robert Michael Pyle, PhD, author of *Wintergreen* and
> *Nature Matrix*

"Thor Hanson is not just a scientist and writer—he is a gifted raconteur, filled with wonder and love for the wild earth. In *Hurricane Lizards and Plastic Squid*, Hanson brings his unique perspective to this time of ecological crisis. Rather than just a warming planet, we find stories from the infinite and varied tangle of life, with every being—from bacteria to birds—seeking to adapt with ingenuity and resilience. This book bears witness to the individual stories so often lost in climate headlines, and invites us all to live with greater depth and awareness as we seek a hopeful path forward."

— Lyanda Lynn Haupt, author of *Rooted* and *Mozart's Starling*

Hurricane
Lizards
and
Plastic Squid

Also by Thor Hanson

The Impenetrable Forest

Feathers

The Triumph of Seeds

Buzz

For Children

Bartholomew Quill

Hurricane Lizards and Plastic Squid

The Fraught and Fascinating
Biology of Climate Change

Thor Hanson

BASIC BOOKS
New York

Basic Books
Hachette Book Group
1290 Avenue of the Americas, New York, NY 10104
www.basicbooks.com

Printed in the United States of America

First Edition: September 2021

Published by Basic Books, an imprint of Perseus Books, LLC, a subsidiary of Hachette Book Group, Inc. The Basic Books name and logo is a trademark of the Hachette Book Group.

The Hachette Speakers Bureau provides a wide range of authors for speaking events. To find out more, go to www.hachettespeakersbureau.com or call (866) 376-6591.

The publisher is not responsible for websites (or their content) that are not owned by the publisher.

Print book interior design by Trish Wilkinson.

Lizard illustration copyright © by Colin Donihue.

Library of Congress Cataloging-in-Publication Data
Names: Hanson, Thor, author.
Title: Hurricane lizards and plastic squid : the fraught and fascinating biology of climate change / Thor Hanson.
Description: First edition. | New York : Basic Books, 2021. | Includes bibliographical references and index.
Identifiers: LCCN 2021005928 | ISBN 9781541672420 (hardcover) | ISBN 9781541672413 (ebook)
Subjects: LCSH: Bioclimatology. | Adaptation (Biology) | Biotic communities. | Global environmental change. | Climatic changes.
Classification: LCC QH543 .H36 2021 | DDC 577.2/2—dc23
LC record available at https://lccn.loc.gov/2021005928

ISBNs: 9781541672420 (hardcover); 9781541672413 (ebook)

LSC-C

Printing 1, 2021

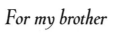

For my brother

Contents

Author's Note   ix

Introduction: Thinking About It   xi

Part 1: The Culprits (Change and Carbon)   1

CHAPTER ONE   Nothing Stays the Same   3
CHAPTER TWO   Mephitic Air   17

Part 2: The Challenges (and Opportunities)   29

CHAPTER THREE   Right Place, Wrong Time   31
CHAPTER FOUR   The Nth Degree   45
CHAPTER FIVE   Strange Bedfellows   57
CHAPTER SIX   The Bare Necessities   69

Part 3: The Responses   83

CHAPTER SEVEN   Move   85
CHAPTER EIGHT   Adapt   101
CHAPTER NINE   Evolve   117
CHAPTER TEN   Take Refuge   135

Part 4: The Results   151

CHAPTER ELEVEN   Pushing the Envelope   153
CHAPTER TWELVE   Surprise, Surprise   173
CHAPTER THIRTEEN   That Was Then, This Is Now   189

CONCLUSION   Everything You Can   207

Acknowledgments   215
Glossary   217
Notes   221
Bibliography   247
Index   267

Author's Note

This is a book driven by curiosity and told through the stories and discoveries of scientists, an inherently curious group of people. Though rooted in the climate change crisis, it is not a crisis book. Other volumes have raised the alarm, and those warnings stand. Here the focus is on underpinnings—how biology teaches us what to expect, when expecting climate change. It is filled with dispatches from the front lines of a rapidly expanding field, and the bibliography contains even more fodder for exploration. I've tried to distill scientific ideas without too much jargon, but there is a glossary in the back for the unavoidable terms that slipped in. Anecdotes and asides that fell outside of the narrative are included in the chapter notes, including details on building a better beetle trap, the longevity of packrat urine, and how to dissolve a duck egg in water. I hope that the many insights I've gained in researching and writing this book will be mirrored in the reading of it, and that it sparks a desire to take action as well as interest. Shouting from the rooftops carries farther when we all raise our voices together.

Introduction

Thinking About It

I am thinking, brother, of a prediction I read . . .
—William Shakespeare
King Lear (c. 1606)

I pitched my tent in the dark and the pouring rain, hoping I'd scrambled far enough up the slope to be out of the range of flash floods. Crawling inside was like entering a washing machine on spin cycle—wind lashed the wet fabric inches from my up-turned face, rattling the tent poles and spraying me with a fine mist. As the storm raged late into the night, and as my sleeping bag slowly soaked through, I began to second-guess my choice of activities for the spring break holiday.

I could have joined friends on a fishing trip, partaking in the sort of beery camaraderie that is more or less expected of college students during the final term of their final year. Instead, I de-cided at the last minute to make a stack of sandwiches, throw my camping gear into a backpack, and head out to explore a re-mote corner of the Southern California desert that would one day become Joshua Tree National Park. It never occurred to me to pack waterproof tarps and rain gear—I was going to the driest place in North America! But while that first night was among the most miserable I've ever spent in a tent, its rain produced a wondrous result. Thirsty seeds and perennials sprang to life all around, and as the skies cleared in the days ahead I found my-self hiking through that rarest of landscapes—a desert in bloom. My field notes describe a profusion of gold, blue, and purple blos-soms, splashed like brushstrokes across the red earth and granite. I recorded over two dozen species in flower, from bright daisies and bluebells to less familiar varieties with names straight out of a Western novel: scorpion-weed, Spanish needle, and jackass clover. The plant that I wrote about most, however, didn't have flowers at all. It bore decorations of a different kind.

I came across it growing alone in a narrow mountain pass, an old Joshua tree with branches that spread upward like the tines of a rake. Even from a distance, I could see that it shimmered oddly as it swayed in the breeze, and when I got close I knew the reason. Prevailing winds, channeled by rocks and elevation, had festooned the tree with trash. There were plastic bags, food wrappers, strands of baling twine, and no fewer than three helium party balloons in varying stages of deflation. "Happy Birthday," one still read, shaking feebly at the end of its tangled ribbon. At the time, I compared the litter to fruit—a strange harvest so deep in the wilderness, fifty miles from the nearest sizable town. De-cades later, I can still picture that tree and it still strikes me as a potent symbol for our far-reaching impacts on the natural world. But I recognize now that the problem wasn't so much in what the windy air had deposited; it was in the air itself.

FIGURE I.1. The Joshua tree is the world's largest variety of yucca and grows exclusively in the Mojave Desert, a region changing rapidly as the climate warms. National Park Service / Robb Hannawacker.

Two months after that hike, I collected my undergraduate diploma and began a career in conservation biology. By chance, my graduation day occurred just as delegates were gathering for the 1992 Earth Summit in Rio de Janeiro, Brazil, where they would introduce and sign the first international treaty on climate change. It wasn't a new concept—scientists predicted the impact of carbon emissions in the nineteenth century, and the phrase "global warming" had been common in environmental circles for years. But the Earth Summit marked a turning point, the moment when climate change officially transitioned from a scholarly topic to a global public concern. In the years ahead, mounting evidence and calls for action would clash repeatedly with politics, particularly in the United States. There would be climate change protests, campaigns, and debates, not to mention that ultimate sign of collective angst: a string of Hollywood disaster movies. As a scientist, I never doubted the urgency of the issue, but I still struggled alongside everyone else to find a meaningful response. The irony of flying to far-flung field sites in places like Africa and Alaska did not escape me—I wasn't exactly going to cancel out burning all that jet fuel by carpooling to the airport. But beyond such hazy worries, the climate problem felt remote at first, alarming but intangible, like a diagnosis in want of a symptom.

My reaction was typical. When it comes to climate change, there is a glaring disconnect between what we know is happening and what we seem able or willing to do about it. Longtime climate campaigner George Marshall explored this disparity in his excellent and aptly titled book *Don't Even Think About It*. He noted how the human brain is perfectly capable of simultaneously understanding and ignoring abstract threats. When consequences seem distant or gradual, the rational part of our mind simply files them away for future reference and rarely triggers the more instinctive, emotional pathways associated with quick action. (We do better responding to physical threats, such as spear thrusts and charging lions, the sorts of immediate problems that our ancestors evolved with.) Marshall's book ends with a laundry list of strategies for bridging that mental gap, many of which rely on something else the human brain is known for: storytelling.

When complex ideas are attached to a narrative, they immediately become more relatable. There is a reason why Plato framed so many of his philosophical dialogues around the drama of the trial of Socrates, and why Carl Sagan chose to teach astrophysics from the glowing deck of an imaginary spaceship. Stories engage parts of the brain left untouched by facts alone, releasing chemicals that demonstrably change the way we think, feel, and remember. Learning about climate change is no different, and much of how we understand and act upon it will ultimately boil down to stories—those we tell, and, in another sense, those that it tells to us. My own perspective has shifted dramatically over the course of my career, transformed from detachment to utter fascination by narratives—not necessarily the ones found in headlines or policy debates, but by those playing out in some place more fundamental: the lives of the plants and animals I've studied.

Like biologists everywhere, I've watched climate change leap from background to forefront in project after project, because while people may have spent the past thirty years struggling to even think about a response, every other species on the planet has simply been getting on with it. Their reactions remind us that the outcome of every future climate scenario, no matter how complex or contentious, relies ultimately on one thing: how individual plants and animals respond to change. If every living thing on Earth got along just as well in any situation, then tweaking the weather wouldn't matter in the slightest. Conditions for life, however, are anything but universal. Biodiversity stems from specialization—millions of species intimately adapted to the nuances of their own particular niche. Altering those conditions forces a response, and when that alteration comes quickly it can restructure whole ecosystems. The speed of climate change is a large part of what makes it a crisis. But for scientists, farmers, birdwatchers, gardeners, backyard naturalists, and anyone with an interest in nature, it also creates an opportunity. Never before have people been in a position to witness such a radical biological event, and if the early results are any indication, it has a great deal to teach us. Because just as the planet is changing faster than anyone expected, so too are the plants and animals that call it home.

This book is an exploration of that emerging world, where species from beetles to barnacles (and even Joshua trees) are meeting the challenge of rapid change head-on—adjusting, adapting, and sometimes measurably evolving, all in real time. Apart from a brief introduction to carbon dioxide, this book does not include detailed explanations about why and how the planet is warming; nor does it address the many controversies that continue to hamper progress on policy. Those are vital topics, but they have been extensively covered in the press and elsewhere. (For an excellent summary, I refer readers to Andrew Dessler's lucid and even-handed text, *Introduction to Modern Climate Change*.) Instead, this book delves into what some are calling a distinct new field of study—climate change biology. Beginning with chapters about how scientists discovered that the climate was changing, and that greenhouse gasses were the culprit, the narrative then follows three questions at the heart of this emerging field: (1) What challenges does climate change create for plants and animals? (2) How do individuals respond? and (3) What can the sum of those responses tell us about the future—theirs as well as our own?

In reading this book I hope you will come to agree with me that climate change deserves our curiosity as well as our concern. It's hard to solve a problem if we aren't even interested in it. Fortunately, this is a crisis that happens to be deeply and profoundly fascinating, affecting the biology of the world around us in ways worth thinking about every day. I am writing these words, for instance, on a fine spring afternoon, with my office door flung wide open to the buzzing of insects in the orchard and the trill of warblers newly arrived from points south. Rising global temperatures touch every aspect of this scene, from the pace of pollination and migration to the fact that my door is open and that I'm

comfortable wearing a short-sleeved shirt. Understanding biological responses to climate change can help us find our place within
it, and it's my hope that the stories in this book will inspire as well
as inform. Simply put, if bush crickets, bumblebees, and butterflies can learn to modify their behaviors, then it stands to reason
that we can too. Plants and animals have a great deal to tell us
about the nature of what comes next, because for many of them,
and also for many of us, that world is already here.

PART ONE

The Culprits
(Change and Carbon)

If you want to make enemies,
try to change something.

—Woodrow Wilson
Address at the Salesmanship Congress (1916)

A s a prospective graduate student, I spent months look-ing for the right doctoral program—touring various university campuses, writing emails, talking on the phone, and meeting with potential advisors. I knew I'd found the right match when I interviewed with a professor who didn't bother showing me his lab or office until after we'd spent a day to-gether out in the woods. "Let's go for a walk," he said, "and see if we have anything to talk about." It was a lesson in the importance of fundamentals, making sure that the basics are covered before getting too far into a complex endeavor. With that in mind, the first chapters of this book focus on essentials that often get glossed over: how scientists started thinking about change and carbon dioxide in the first place . . .

CHAPTER ONE

Nothing Stays the Same

All change in habits of life and of thought is irksome.

—Thorstein Veblen
The Theory of the Leisure Class (1899)

I heard them before I saw them, screeching and croaking from somewhere overhead like a pair of deranged roosters. The noise went on and on, and it struck me as preposterous that any sane person would want to keep one of these birds inside their house. Yet demand for the pet trade had helped transform the great green macaw from a commonplace species into an endangered species. I'd spent three years studying their main food source in what was once prime habitat, but to actually spot a macaw required two days of backcountry travel by bus, river launch, and finally, a motorized canoe. So when two birds suddenly launched themselves from a treetop and soared out over the river, I felt the thrill of a moment long anticipated, and I also knew immediately what made pet fanciers so willing to overlook all that racket and clamor. Even from a distance, the macaws' brilliant green plumage shone in the sunlight, rippling with accents of crimson, chestnut, and bronze, and framed by wide blue wings, as if every color within view, from sky to river to rainforest, had been distilled and brought to life in feathers.

I watched with satisfaction as the birds crossed from the Nicaraguan to the Costa Rican side of the river and disappeared over a row of low hills. It seemed fitting to close out my research in Central America by glimpsing evidence of the avian resettlement it was designed to encourage. Though I hadn't studied macaws directly, my work showed that almendro trees—whose almond-like nuts the birds rely upon—could persist and reproduce in patches of forest indefinitely, connected to one another over long distances by the busy pollination efforts of bees. That finding helped justify a new law protecting almendros throughout the lowlands of eastern Costa Rica, where cattle ranching and fruit production had left the rainforest divided by pastures, roads, and cropland. People hoped that if the right kind of trees remained, the macaws might return, repopulating old haunts from their stronghold to the north, in the large Nicaraguan nature reserve that I'd traveled so far to visit. As it turned out, that process was already well under way. In the years ahead, hundreds of birds would set off on the same flight I'd witnessed, crossing the San Juan River and heading south to once again make great green macaws a regular sight (and sound) in parts of Costa Rica. It was briefly held up as a conservation success story—the returning birds not only found food in almendros, they also nested and raised chicks in hollows within the trees' massive trunks. But scientists soon realized that the fate of the macaws and their favorite tree was an even better example of something entirely different, and far more consequential.

Looking back, I see now that the phrase "climate change" did not make a single appearance in the many proposals, reports, and peer-reviewed papers associated with my almendro research. At the time, it didn't seem relevant to such a specific and local biological study. But I did receive one suggestive hint along the way, delivered in an offhand comment from another scientist working out of the same field station. Her data showed how almendro trees responded to hot weather by increasing their rate of respiration, the process plants use to get oxygen to their cells. In a sense, the trees were panting. This and other signs of stress didn't bode well in a warming world, and when climate modelers later began making predictions about Central America, it was clear that almendros were in a tight spot. "The trees you studied will be gone by the end of the century," one expert told me, explaining how the species' survival depended on shifting its range upward in elevation to find the temperatures it preferred. Suddenly, the most important result of my work was something I'd published almost as an afterthought—the fact that large fruit bats could disperse almendro seeds in leaps of a half mile (eight hundred meters) or more. Would that be far enough and fast enough to beat the heat? Would the bats be moving in the right direction? Could almendros even establish themselves in higher forests already crowded with trees? And what did all this mean for the macaws, who were expected to simply fly north to cooler climes, unconstrained by the slow pace of seed dispersal? Instead of a tidy relationship between parrots and trees, the macaw-almendro story has become yet another case study in uncertainty, symbolic of a planet in flux.

As a biologist, perhaps I shouldn't have been surprised by the sudden plight of almendro trees. After all, change lies at the heart of evolution, and evolution is the heart of biology. The very word *evolve* comes from a Latin verb meaning "to unroll," and every organism is a product of that constant motion. Species wheel into existence, adapting and often giving rise to new things along the way, before eventually winking out as the world moves on around them. Even if almendros fail to reach the foothills and disappear altogether, that would be perfectly normal; extinction is the fate of all species. I knew this, but still found it head-spinning to think

FIGURE 1.1. The great green macaw is the largest parrot in Central America, where its relationship with almendro trees is now uncertain. P. W. M. Trap, *Onze Vogels in Huis en Tuin* (1869). Biodiversity Heritage Library.

that my giant study trees—some measuring ten feet (three meters) in diameter—might soon be gone. It was more than sentimentality or simple surprise. Resistance to change is considered a hallmark of the human psyche. Experts link it to our instinctive sense of comfort and safety in the familiar, combined with a need for social cohesion and consistency. The result is a common sentiment neatly captured in the words of cartoon everyman Homer Simpson: "No new crap!"

I certainly wasn't the first person unsettled by the idea of a changing environment. For most of human history, people preferred to dismiss the notion entirely and regard the natural world as something immutable. Certainly, there were seasons and the occasional drought or flood, but the land and the seas and the creatures within them were fixed. Greek philosopher Parmenides went so far as to prove that change was impossible. Nothing comes from nothing, he argued, nor can anything come from what already exists, because, "what is . . . *is*."

Aristotle found some wiggle room in that argument by suggesting that objects might change form so long as their underlying essence persisted. An acorn could grow into an oak tree, for example, or bronze could be melted and cast to form a statue. This accounted for the obvious processes of change encountered in daily life, without challenging the idea of nature as something absolute. Aristotle also proposed organizing the natural world into a strict hierarchy, with what he perceived as simpler forms like plants near the bottom and more sophisticated things like animals (and Greek philosophers) on top.

Later scholars embraced and embellished this notion, finding rungs on the ladder for any newly discovered species, as well as things like precious metals, planets, stars, and even various types of angels. The paradigm held for nearly two thousand years, and it was echoed in the taxonomic ranking system developed by that great cataloger, Carl Linnaeus, who noted in 1737 that all true species "have been assigned by Nature fixed limits, beyond which they cannot go," and that their number "is now and always will be exactly the same." Even as Linnaeus wrote those words, however, new ideas were already shaking the foundations of the old worldview. Fittingly, the evidence that change was not only common, but in fact a prime mover in nature, came from stone, a substance that had always been placed at the very bottom of Aristotle's hierarchy.

Few readers are thought to have made it through all 1,548 pages of James Hutton's 1795 opus, *Theory of the Earth*, not to mention

FIGURE 1.2. This sixteenth-century illustration depicts the natural world as an immutable "Great Chain of Being," ascending from rock and soil to plants, animals, and humanity. Images of heaven and hell (and their inhabitants) frame the scene above and below. Diego Valadés, *Rhetorica Christiana* (1579). Getty Research Institute.

its 2,193-page companion, *Principles of Knowledge*. But even the Scotsman's daunting wordiness couldn't obscure the power of his central geological theme—that the bedrock of continents and islands formed from constant erosion and sedimentation, cemented and then uplifted by the heat of the Earth. Instead of a static landscape, he proposed an ongoing "succession of worlds," continually unfolding over huge spans of time. It was a radical thought, but one supported by ample evidence then coming to light in the mine shafts proliferating across Great Britain. Demand for coal and metals to feed the Industrial Revolution had inadvertently opened a window into deep time, exposing layers of bedrock with ancient stories to tell. Some contained marine fossils, bolstering Hutton's notion that rocks—even those found high up on hills and mountains—had formed from ocean sediments. Other stones held the remnants of strange plants or unfamiliar animals, suggesting that life, as well as landscapes, had looked quite different in the distant past. This raised an obvious and troubling question: Where had those species gone?

Extinction was a purely hypothetical concept until French naturalist Georges Cuvier started thinking about elephants. Shortly after Hutton upended the idea of permanence in geology, Cuvier took aim at its biological counterpart. His meticulous examination of fossil elephant teeth showed that various mastodons and woolly mammoths were distinctly different—not only from one another, but from all living elephant varieties. He called them lost species, and because elephants are enormous and impossible to overlook, it was hard for doubters to argue that mammoths and mastodons were still out there somewhere, waiting to be noticed. (Interestingly, mastodon enthusiast and third US president Thomas Jefferson suggested just that, instructing members of the 1804 Lewis and Clark Expedition to scour the American West for animals that "may be deemed rare or extinct.") Cuvier spent the rest of his career driving the point home, describing extinct forms of everything from turtles and sloths to pterodactyls. But one of his most lasting contributions was the observation that species didn't just wink out one by one. Sometimes whole communities disappeared from the fossil record all at once, replaced by a vastly different group of organisms in shallower, younger layers of rock. He famously held this up as a challenge to Hutton's ideas about gradual geological change, arguing that ancient landscapes (and all their inhabitants) had instead been repeatedly destroyed by a series of floods or other catastrophes. As a general theory, known as catastrophism, it was eventually debunked. Aside from the occasional earthquake or volcano, most processes in geology do indeed play out slowly, just as Hutton had suggested. But Cuvier's fossils showed that extinction events could at least occasionally be abrupt and widespread—the first indication that the natural world was capable of rapid change. It was an idea that the greatest naturalist of the next generation would always struggle to reconcile.

Hutton's and Cuvier's theories challenged religious norms as well as scientific dogma, and decades of contention followed. Many scholars countered with biblical arguments—if rocks contained traces of marine life, then they must have formed during the Flood, and any unfamiliar fossils were simply creatures that hadn't made it onto the ark. Others accepted the concept of ancient worlds, but offered different theories about rock formation, fossil origins, and what caused the transition from one era to the next. Such debates fascinated the young Charles Darwin, who devoted much of his early career to geology. He called himself a "zealous disciple" of the Hutton viewpoint, as popularized and expanded upon by the great nineteenth-century geologist (and Darwin's good friend) Charles Lyell. Darwin collected thousands of fossils and rock specimens during his voyage on the Beagle— often at the expense of zoological pursuits—and looked forward

FIGURE 1.3. In *Exhuming the First American Mastodon*, artist and naturalist Charles Willson Peale immortalized his own 1801 excavation of a creature originally dubbed the American incognitum. Sketches of the fossil eventually reached Georges Cuvier in Paris, who confirmed it as a mastodon, one of the first species definitively established as extinct. Maryland Historical Society.

to visiting the Galápagos Islands not for their finches, but because "They abound with active Volcanoes." He later drew on fossil evidence to support his thinking about species formation, and so did Alfred Russel Wallace. Joint publication of their papers on evolution by natural selection in 1858 (and Darwin's *The Origin of Species* the following year) did for biology what Hutton had done for geology—embracing change as fundamental, and giving it a convincing mechanism. But both men considered the pace of that change to be slow and incremental, neatly complementary to the emerging consensus on gradual geological forces like erosion and sedimentation. More than a century would pass before biologists began to grasp how quickly things could happen—in the environment, in evolution, and in the critical ways those forces interact. Once again, the first insights came not from studying modern creatures, but from an understanding of stone, fossils, and vast spans of time.

In 1971, two newly minted paleontologists introduced the phrase "punctuated equilibrium" at the annual meeting of the Geological Society of America. Friends and collaborators since graduate school, Niles Eldredge and Stephen Jay Gould presented their idea as a novel answer to a question that had long plagued the field of paleontology: Where were the missing links? If evolution was indeed a slow and steady process, then shouldn't the fossil record be filled with gradual transitions from one form to the next? Instead, fossil species tended to appear abruptly and then persist, more or less unchanged, through layers of rock representing thousands or even millions of years. Darwin had been well aware of this problem, calling it "the most obvious and gravest objection which can be urged against my theory." He devoted a full chapter in *The Origin of Species* to an explanation that people had relied upon ever since: "the extreme imperfection of the geological record." Because rocks form only under the right conditions, and only a tiny fraction of rocks contain fossils, the vast majority of species (and the transitions from one to another) have gone unrecorded. In Darwin's memorable description, "I look at the natural geological record, as a history of the world imperfectly kept . . . only here and there a short chapter has been preserved; and of each page, only here and there a few lines." Eldredge and Gould didn't dispute the limits of the geological record, but they suggested something else that would make transitional fossils rare: rapid evolution. If new species arose in quick bursts of change instead of slowly emerging over eons, then there simply wouldn't be time—from a geological perspective—for their transformations to leave a trace.

Punctuated equilibrium managed to challenge evolutionary thinking without actually challenging evolution—natural selection and all the other basic Darwinian principles still applied. Only the tempo was different. A process that involved bursts of rapid activity (punctuations) followed by long periods of stability (equilibria) could explain the fossil records of everything from trilobites to horses, and supporters began applying it broadly. Critics accused Eldredge and Gould of overstating or misinterpreting their case, exaggerating what might be only a minor trend of fits and starts in an otherwise gradual system. That debate continues, but regardless of whether the pattern is common or rare, or exactly what causes it, punctuated equilibrium introduced an important idea: that the rate of evolutionary change is variable, and that—at least some of the time—it moves in rapid bursts.

Over the course of two centuries, scientific and popular perceptions of nature went from something fixed and inviolate, to something that changes slowly in tiny steps, to something capable of swift and abrupt transformations. Biologists saw their role expand accordingly. Instead of simply cataloging species, they began decoding their histories and relationships, and began looking for measurable signs of evolution in action. How did plants and animals respond to their environment, and to one another? What made some species resilient enough to persist for millions of years, while others (like almendro trees) seemed vulnerable to the slightest perturbation? What conditions led to pulses of activity—in the evolution of species as well as the rate of extinction? All of these questions played out against another growing realization. In study after study, in ecosystems around the globe, one species kept emerging as the dominant and overarching agent of change.

Traditional views of nature did not include a significant role for the impacts of human behavior. Farming, hunting, logging, and other activities may have exacted a toll, but these costs were seen as local and temporary. When the Roman emperor Trajan's victory over Dacia was commemorated on a column, for example, the bas-relief carvings showed a wooded kingdom being denuded and stripped of wildlife to supply the conquering army. But it was implicit that this rich landscape would soon recover—why else would Dacia have been worth conquering? In the words of an old Chinese saying, "As long as green hills remain, there will always be firewood." It wasn't until well into the nineteenth century that people began to realize those proverbial hills were less than inexhaustible. Industrialization, urbanization, and population growth all brought environmental consequences that people could experience firsthand, from air and water pollution to shortages of game, arable land, and yes, firewood. Overhunting had settled the extinction question once and for all, eliminating common species like the passenger pigeon and great auk, as well as high-profile exotics like the dodo. When German naturalist and explorer Alexander von Humboldt warned in 1819 that cutting forests would create "calamities for future generations," most people were still skeptical. But by the end of the century, governments around the world had begun setting aside parks, forest reserves, and wildlife refuges as a matter of course, and a growing network of citizen groups were lobbying to protect the environment. It was another of Von Humboldt's insights, however, that hinted at our current predicament, when he suggested that "vast amounts of gas and steam" given off by centers of industry were altering the climate.

To be clear, Von Humboldt viewed factory emissions as a strictly local concern, something that threatened to trap heat in and around large cities. He believed that broader climate trends depended on aspects of geography, prevailing winds, and other factors "upon which civilization has no significant influence." But as industrialization progressed and air pollution intensified,

more and more people began thinking about the scale of its consequences. Downwind health effects inspired "smoke abatement"
societies across Europe and North America, and in the 1850s, a
study of the notorious murk surrounding Manchester, England,
established that burning high-sulfur coal caused acid rain. At the
same time, physicists had determined the heat-absorbing capacity
of water vapor and various gasses, confirming their role in regu-
lating atmospheric temperatures. A few decades later, Swedish
chemist, physicist, and Nobel laureate Svante Arrhenius brought
all these threads together, suggesting that humanity's "consumption of coal, petroleum, etc." was indeed capable of changing the
climate—not just locally, but for the entire planet. He predicted
that "any doubling of the percentage of carbon dioxide in the
air would raise the temperature of the earth's surface by 4°." But
whether out of some sense of optimism, or a basic faith in the
human endeavor, or simply because he lived in chilly Sweden,
Arrhenius thought this increase in temperature sounded like a
great idea. Human-induced climate change would lead to better
weather and higher crop yields, he argued, and help stave off the
possibility of another ice age.

Few people took note when Arrhenius published his climate
predictions in 1896, and more than a half century passed before
there was equipment accurate enough to test and refine them. But
as carbon dioxide levels and global temperatures began measurably rising together, the basics of the Arrhenius premise became
a cornerstone of climate science. Modern practitioners don't exactly agree with his rosy outlook on the consequences, however,
and there is another aspect of climate change that the visionary
Swede certainly got wrong: its speed. Presenting his findings at a
public forum in Stockholm, Arrhenius told the audience that human activity was on track to double atmospheric carbon dioxide
in three thousand years. If current emissions hold steady, we will

reach that milestone in less than thirty. Once again, the planet's capacity for change is exceeding our expectations, prompting scientists in the twenty-first century to ask not whether sudden transformations are possible, but whether we are living within one.

In the history of thinking about nature, the concept of rapid change still ranks as a relatively new idea. This helps explain why the present moment is so pivotal, and so full of surprises. Modern climate change is transforming theoretical abstractions into sudden realities, putting many of the processes that shaped life and landscapes during past global upheavals on full display. Because this book explores how species respond, there will be little discussion of the complexities (and controversies) of causation. After all, plants and animals aren't concerned about *why* the planet is warming; even if this was a natural trend, their predicament would be the same. But there is one climate change culprit—often invoked but rarely explained—that requires further investigation.

As a field scientist, I'm used to studying things that I can see. I will happily invest days of travel time for a glimpse of rare parrots flying across a river, because I've learned that direct observations always help me to think, understand, and ask better questions. It goes without saying that the consequences of climate change are now very evident in nature—that is the basis for this book. But the driving force behind it remains invisible—which invites a fundamental but often overlooked question: Just what, exactly, is carbon dioxide? And, for that matter, where can I get my hands on some?

CHAPTER TWO

Mephitic Air

*We must measure all that is measurable, and strive to
make measurable all that is not . . .*

> —attributed to Galileo
> by Thomas-Henri Martin
> *Galilée* (1868)

The illustration of a carbon dioxide molecule in my high school chemistry textbook showed a large black ball (the carbon) bookended by two smallish red ones (the oxygen). At the time, I recall thinking that it looked just like the face of a red-eyed fruit fly, the species we'd spent the previous semester studying in biology. Sketch on some mandibles and a pair of antennae, and it made a perfect headshot! That association stuck with me, and later, when carbon dioxide's link with climate change became notorious, I found myself picturing the world's tailpipes and smokestacks spewing forth an endless swarm of tiny flies. It was a vivid image, but didn't really tell me much about the gas in question. More persistent and abundant than methane or the other greenhouse contributors, carbon dioxide is at once ominous and essential, a global threat that also happens to be one of the building blocks for life on Earth. That ubiquity makes it relatively easy to find, which helps explain why it was the first atmospheric gas to ever be identified. In fact, before the discovery of carbon dioxide, scientists weren't sure exactly what the atmosphere was, or whether it contained anything measurable at all.

In the summer of 1767, the noted English theologian, natural philosopher, and all-around polymath Joseph Priestley had some time on his hands. His light duties as a minister in Leeds left most of his days free, and he used that time to think, write, and tinker. Having already produced books and papers on everything from grammar to electricity, he chose as his next topic an exciting new field known then as pneumatic chemistry, the study of gasses. That decision sparked what one biographer has called "an intellectual streak of legendary proportions." Within a few short years, Priestley would firmly establish not only that air was measurable but that it was complicated—a swirling mixture of distinct components. Along the way, he became the first person to isolate and describe oxygen and ten other common gasses, not to mention uncovering the basic chemistry behind photosynthesis. But it all started with his curiosity about something miners called choke damp, or, more poetically, mephitic air, an invisible, suffocating vapor known to gather at the bottom of coal shafts. Scottish chemist Joseph Black had recently cooked up a batch of the stuff in his laboratory by heating small pieces of chalk and limestone and trapping the fumes in a bottle. Luckily for Priestley, there was another place this gas was known to occur, and he happened to live right next door.

"I was induced to make experiments," he later reminisced, "in consequence of living for some time in the neighborhood of a public brewery." There, lingering over the vats of fermenting ale, Priestley found a ready supply of the gas in question, "generally about nine inches, or a foot in depth, within which any kind of

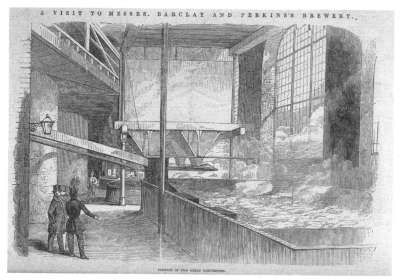

FIGURE 2.1. The fermentation of beer produces large amounts of carbon dioxide as a by-product, giving Joseph Priestley a living laboratory for his gas experiments. Barclay and Perkins brewery (1847). Wellcome Collection.

substance may be very conveniently placed." Over the coming months he placed a remarkable array of things into that bubbling zone: candles, hot fire pokers, ice, rosin, sulfur, ether, wine, butterflies, snails, mint sprigs, various flowers, and at least one "large, strong frog." Perhaps the only thing more boundless than Priestley's curiosity was the patience of the brewers, who indulged their eccentric vicar even when his experiments went awry and left the beer with "a peculiar taste." Like previous observers, Priestley immediately noted how the gas seemed to lack something. Candle flames were snuffed out in it, and animals left for any length of time quickly succumbed to asphyxiation. (Happily, the "strong frog" was rescued and revived after only a few minutes.) But Priestley also recognized that this mysterious vapor was more than simply an absence of "normal" air—it boasted unusual qualities that made it quite distinct and interesting in its own right. He learned that it would bleach the color from rose petals. He saw that it was heavy, and he watched smoke become trapped within it and stream down the sides of the vats to gather on the brewery floor. Most famously, he discovered how to quickly dissolve the gas in water, producing a fizzy beverage with a "pleasant acidulous taste." This breakthrough earned Priestley the prestigious Copley Medal from the fellows of the Royal Society of London. It earned a lot more for entrepreneur Johann Schweppe, who copied Priestley's methods to found the tonic and soda water company that still bears his name. Thanks to such early advances, people knew quite a lot about mephitic air—including its flavor—long before chemists came up with the name carbon dioxide.

FIGURE 2.2. Credit for the discovery of carbonated water goes to Joseph Priestley, but it was Johann Schweppe who saw its vast economic potential. Advertisement (1883). The British Library.

Nearly two and a half centuries after its publication, Priestley's book on gasses still brims with excitement. Reading it on a blustery December morning, I found myself caught up by his enthusiasm, eager to experience those same revelations that he had at the brewery. My son, Noah, was a willing accomplice. He happened to be home from elementary school for the day, suffering from just the right sort of head cold—sick enough to miss classes, but healthy enough to enjoy the time off. "Let's find some carbon dioxide!" I told him, and the game was afoot.

We could have simply popped open a few soft drinks and tried to collect the bubbles. (Appropriately, there were even a few bottles of Schweppes on the shelf.) But somehow, relying on carbonation to get our carbon dioxide seemed a bit too much like cheating. The smoke from our blazing woodstove certainly contained the desired vapors, but how would we filter out the sixty or more other gasses, chemicals, and nasty particulates? Instead, it seemed best to follow Priestley's example and go to one of the purest and most common sources of carbon dioxide on Earth. So we headed for the refrigerator.

It turns out that fermentation occurs in a lot more places than vats of beer. Yogurt and cheese makers call it culturing, but it's more accurate to think of it as a form of slow microbial digestion, a way for bacteria and other tiny organisms to extract and use energy from the foods they live within and around. Like any form of digestion, it is a process that produces waste. Luckily for food lovers, the by-products of fermentation include things like alcohol (thus the beer) and lactic acid, which adds tanginess and pungency to such cultured favorites as kimchi and buttermilk. Most fermentation also produces carbon dioxide, which is what had me combing through the back reaches of our fridge. A container of organic sauerkraut advertised "Probiotic Punch!" and claimed, "It's Alive!" But any organisms the kraut might have contained had long since shuffled off their mortal coils, and were no longer producing carbon dioxide—a lit match held above the briny mix burned brightly. Experiments with yogurt and sour cream were similarly disappointing. But then we hit the jackpot.

On the bottom shelf, behind the bags of carrots and celery, sat a half-gallon jar of homemade pickles. They had been stewing in their own juices since August, and tasted yeasty as well as sour, suggesting that fungi had joined the bacteria in their digestive pursuits. Frankly, it was past time to throw the pickles out, but for once, procrastinating a chore had paid off. As soon as Noah and I brought a match near that open lid, the flame demonstrated why carbon dioxide is such a common ingredient in fire extinguishers. With no oxygen to burn, the match went out in an instant, as if we'd turned off a switch. What's more, smoke from the snuffed tip curled *downward*, trapped in the gas just as Priestley had described.

"It's pouring down the side!" Noah exclaimed, watching as wisps of smoke followed the heavy vapors over the jar rim and down to spread across the countertop.

"That's it," I told him. "You saw the carbon dioxide!"

He quickly reminded me that our quarry was invisible. "I didn't see the carbon dioxide, Papa. I saw the smoke." But like Priestley before us, we could use that smoke to watch the gas, defining its boundaries as it flowed and swirled around the open jar. For a few minutes our kitchen was filled with the thrill of discovery as we lit match after match, watching them snuff and smolder, until all the carbon dioxide had dissipated into the surrounding air.

Simple experiments often lead to broader insights, and repeating Priestley's fermentation trick brought up an obvious question: Do pickles cause climate change? Does making beer? The answer, of course, is no. But understanding why some carbon emissions are harmless, while others are not, reveals a basic truth about climate change that people rarely stop to think about.

In the case of our pickle jar, the carbon came from the cucumbers in the brine, and the cucumbers had gotten it from the air around our garden the previous summer. Like plants everywhere, their growth relied on photosynthesis, that leafy process of combining carbon dioxide and water with energy from the sun to create starches. (In other words, carbon dioxide puts the *carbo-* in carbohydrates.) When those starches break down, the carbon dioxide goes back into the atmosphere. This is the step in the Earth's carbon cycle that we are most familiar with, because we play a role in it every moment of every day. Whether we eat plants, or we eat animals that have eaten plants, the energy that fuels our bodies traces right back to those photosynthetic starches, and we release carbon dioxide with every exhaled breath. But in terms of climate change, breathing is like making pickles or brewing beer—guilt-free. That's because our bodies are just one short stop for carbon on its continuous circuit from the air through plants and animals and back again, with no net gain or loss. If that's all there was to the story, then the planet wouldn't be warming and I wouldn't be writing this book. The reality of modern climate change hinges on one key fact: not all plants break down.

Consider the pickle. Cucumbers eaten fresh or left to rot in the garden release their carbon right away, but that process slows considerably in a jar of salty brine. Under the right conditions, it can stop altogether. In nature, this occurs primarily in two locations: the ocean floor and boggy wetlands. When marine algae die en masse and sink to the seabed, they sometimes get buried before they are eaten or decomposed. Dead plants in bogs can also accumulate with little decay, forming layer upon layer of peat. In either case, if sedimentary rocks develop above and around these organic deposits, their carbon is effectively trapped and removed from the atmosphere for millions of years. Transformed by heat, pressure, and age, these ancient plants are now as familiar to us as the fossil fuels—petroleum (from the algae), coal (from the peat), and natural gas (from either). Burning them returns that stockpiled carbon dioxide to the air all at once, overwhelming the natural cycle, and leading to the many consequences now unfolding.

Academically, I knew these things long before reading about Joseph Priestley's experiments. I also understood that carbon moved through the environment by other means, like erosion and volcanic activity, and that it was locked away in types of limestone formed from sediments rich in shells and coral. (While Priestley's brewery trials were benign for the climate, Joseph Black's experiments involved burning chalk and other forms of limestone, a key step in cement production, which is another way people are putting ancient carbon back into the atmosphere.) But finding a source of carbon dioxide percolating harmlessly in our refrigerator made the whole cycle come to life, and it clarified the distinction between normal everyday sources of carbon and the fossilized ones causing all the trouble. At the end of our experiment, Noah and I carefully sealed the pickle jar and tucked it back into the fridge, hoping it would fill up with gas again soon. There was one more critical piece of evidence that I wanted to see for myself.

Following Priestley's discoveries, and with Johann Schweppe peddling his wares across Europe, it's not surprising that other scientists soon began to investigate this readily available gas. Irish physicist John Tyndall made the next leap forward by discovering that carbon dioxide absorbed radiant heat, the very trait that puts it at the heart of modern climate change. I read his papers on the subject, and quickly realized that duplicating his experiments was out of the question. Tyndall isolated his gas samples in a handcrafted copper and iron tube so ingenious and elegant that it now sits on permanent display at the Royal Institute in London. But while my pickle jar was a crude substitute for Tyndall's famous

FIGURE 2.3. John Tyndall drew crowds to his public lectures in London, and he was known not only for his scientific insights, but for the clever instruments he devised to test them. *London Illustrated News* (1870). Wikimedia Commons.

instrument, the old physicist might have envied my heat source. Where he struggled with finicky metal plates and a cube filled with hot oil, I had the benefit of electricity, and no small experience raising chickens.

Whenever my family orders a new batch of hens they come straight from the hatchery as day-old chicks. (The US Postal Service prohibits mailing live animals but makes specific exceptions for young poultry, as well as honeybees and, rather mysteriously, scorpions.) For the first several weeks the tiny birds reside in our living room. In lieu of a mother chicken to keep them warm, we dangle a heat lamp above the open top of their cardboard box, adjusting its height to get the temperature just right. Too low and the chicks scramble away from the glowing bulb, hot and panting.

Too high and they form a cold huddle directly beneath it. With a bit of tinkering, it's easy to manipulate the climate of the box and keep things perfectly cozy. With a bit more tinkering, this system held great promise for testing the effects of heat on carbon dioxide. The only thing I needed to do was replace the baby chickens with pickle jars.

To be honest, my expectations were pretty low. John Tyndall had spent months inventing and calibrating his equipment, and modern laboratories were even more sophisticated. It seemed absurd to think that the most important (and often contentious) premise of climate change could be so easily reproduced with things lying around the house. But I took what precautions I could, comparing the old pickles from the fridge with a duplicate "control" jar stocked with fresh, nonfermenting cucumbers, and always removing the lids before taking measurements to avoid the competing effects of pressure. (Gasses get warmer under pressure.) After half an hour under the lamp, I measured the temperature in each jar—using four different thermometers, just to be sure—and to my surprise the gassy air above the fermenting pickles was consistently 1.5 degrees Fahrenheit (0.9 degree Celsius) warmer. A few minutes later, after the carbon dioxide had dispersed, the temperatures in the jars were the same. To make sure this wasn't a fluke, I repeated the experiment a few days later (giving the pickle microbes time to make more gas), and got precisely the same result. Like a microcosm of the Earth's atmosphere, the jar infused with extra carbon dioxide reliably trapped and retained more heat than the jar with air alone. The temperature difference was small, but that only reinforced the lesson: when it comes to the climate, modest-seeming tweaks can have dramatic consequences.

Inadvertently, the pickle jar trials did more than provide a hands-on experience with carbon dioxide. They also got me thinking more clearly about the challenges that organisms face during times of rapid change. When I went to repeat the temperature experiment for the third time, I noticed that the pickles lacked their familiar, pungent aroma. Though days had passed since I'd last opened the jar, a lit match waved over the top revealed that not a trace of carbon dioxide had built up inside. Apparently, seesawing repeatedly between the cold of the fridge and the heat of the lamp had proven too much for the microbes in the brine: there was nothing left alive to continue the fermentation process. It was a stark reminder that creatures of all kinds—even salt-hardy bacteria—struggle to cope with an unstable climate. Heat waves, cold snaps, and other extreme weather events have already become hallmarks of modern climate change—not in pickle jars, of course, but in ecosystems around the globe. These events create a wide range of stresses (and a few opportunities), and they mark the perfect starting point for an exploration of climate change biology.

PART TWO

The Challenges (and Opportunities)

Facing it—always facing it—that's the way to get through.

—Joseph Conrad
Typhoon (1902)

E veryone knows how to play checkers. At least, that's what I thought until I sat down to a game with one of my field assistants in rural Costa Rica. Pieces that I expected to only move forward were suddenly jumping in all directions, clearing my side of the board in a matter of minutes. I'd like to blame the loss on my clumsy Spanish, but the fact is I never could beat him, even after I learned the local version of the game. It's hard to adjust old habits and strategies when somebody changes the rules you're used to. The same thing is true in nature, where climate change is altering the playing field for species around the globe. As those baselines shift, plants and animals face four main challenges in keeping up with the action . . .

CHAPTER THREE

Right Place, Wrong Time

We loiter in winter while it is already spring.

> —Henry David Thoreau
> *Walden* (1854)

"You should have been here yesterday," said the woman next to me at the viewpoint. "It was T-shirt weather!"

Looking out across the frozen pond, ringed by winter-bare trees, her claim seemed hard to believe. But it was true: twenty-four hours before I'd arrived in Massachusetts, the thermometer had topped out at sixty-four degrees Fahrenheit (eighteen degrees Celsius), the highest mark on record for early February. Now things were back to normal, hovering around freezing, with a cold wind bringing clouds from the south. It was the kind of day where you had to keep moving to stay warm, so I set off along the trail at a brisk pace, feeling a growing sense of anticipation. For someone who writes books about natural history, walking this particular path amounted to a kind of pilgrimage.

Walden Pond measures less than a half mile (eight hundred meters) across at its widest point, but it occupies a much larger place in the history of environmental literature. Modern

American nature writing practically began there when Henry David Thoreau chose it as the site of his retreat from the hustle and bustle of nineteenth-century society. His 1854 memoir, *Walden*, includes musings about everything from poll taxes to Paris fashions, but most people remember it for its vivid descriptions of the very landscape I was hiking through. If Thoreau could have joined me, he would have no doubt recognized a lot of familiar sights. The neighborhood remains largely rural and wooded, and when I reached the place where his cabin once stood, I found it still surrounded by tall pines and red oak. As an intentional recluse, however, Thoreau might have been surprised by the lack of solitude. Walden Pond now ranks as a global tourist destination, and even on a winter day, the visitors' book contained entries from as far afield as China, Israel, and Belarus. Bus tours from nearby Boston have a dedicated drop-off spot, right next to the gift shop.

The changes at Walden Pond that would have interested Thoreau the most, however, probably have less to do with the people than with the forest he knew so well. That's because his daily routine involved a lot more than thinking, reading, and tending a few beans. He was also a meticulous, almost obsessive, observer of the plants and animals around him. What birds were singing? When did the wildflowers bloom? Which fruits were ripe, and who was eating them? What day did the first leaves appear on the trees, and when did the last ones fall? Thoreau noted all of these things during his long rambles through the woods. And he wrote every single bit of it down.

"It was a gold mine," Richard Primack told me, recalling his first glimpse of Thoreau's data—line after handwritten line of flower observations, laid out by species and date, just like a modern spreadsheet. We were talking in Primack's office at Boston University, a cramped space where overflowing piles of books and

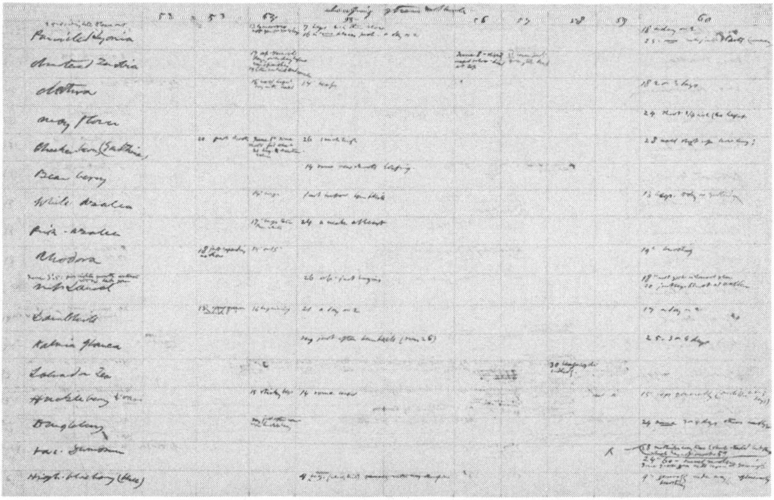

FIGURE 3.1. Henry David Thoreau carefully recorded data on the plants and birds of Walden Pond and the surrounding countryside, laying out his observations by date and species in rows and columns, just like a modern spreadsheet. The Morgan Library & Museum / Art Resource, NY.

papers covered every available surface. The clutter seemed a far cry from Thoreau's doctrine of spare simplicity, but any difference in the two men's housekeeping belied a kindred spirit. "I've seriously considered listing him as a coauthor," Primack said with a laugh. If he had, Thoreau would now rank among the most prolific climate change scientists of the twenty-first century. His specialty, like Primack's, would be phenology, the study of seasonal events in nature. The word is from a Greek phrase for "things that appear," and it carries with it an inherent touch of wonder, as in "phenomenon." The discovery of Thoreau's data was itself a phenomenal and unexpected event, particularly for a specialist in tropical botany.

"The fact is, it was getting harder to work in Borneo," Primack said, citing political and funding issues to explain the abrupt change in his research focus. Still, colleagues were shocked that he would set aside decades of rainforest studies to start poking around in the woods at Walden. "They told me I was crazy, but I saw all these amazing opportunities," he said. In the early 2000s, lots of people were talking about how climate change would impact phenology, but hardly anyone in eastern North America was actually out in the field looking for evidence. Primack began with a single graduate student and a census of wildflowers in springtime. Scores of publications and collaborations later, the project is still picking up steam. "I'm in the most productive period of my career," he told me, with something like bemusement. "At age sixty-nine!"

Nattily dressed in a blazer, and with a fringe of slightly wild gray hair, Primack looks as much like a Thoreau scholar as a botanist, and it's safe to say he's now a little of both. But his initial decision to focus on Walden Pond had little to do with its famous former tenant. He chose it because it was relatively unspoiled, conveniently close to Boston, and well documented by a host of modern naturalists. Primack didn't know about Thoreau's unpublished phenology records—no scientist did. The project was well under way before an offhand comment from a friend in the philosophy department (who happened to be an authority on Thoreauvian ethics) alerted him to the wildflower data, archived at a library in New York. Later, similar serendipity turned up a trove of Thoreau's bird observations, filed in a collection at Harvard. When I spoke with Primack, he was combing through yet another treasure, Thoreau's unfinished book on the seasons, picking out the dates when the first spring leaves appeared on various trees and shrubs. Taken together, Thoreau's records constitute the oldest detailed account of phenology in North America. They are hugely relevant to climate change, because they document not only which species were blooming, budding, or flitting about the woods, but exactly when they did so. That sort of timing lies at the heart of such critical biological events as migration, growth, and reproduction. And when the climate starts warming, it's one of the first things to change.

"Temperature is the overwhelming driver for when plants flower in spring," Primack told me, and it also governs when many species leaf out, and when insects emerge. By combining Thoreau's data with local weather records, and comparing them with more recent observations, Primack and his team showed how flowering times at Walden have advanced by more than four weeks for some species. Violets and wood sorrel that Thoreau admired in May and June now bloom by late April, and what he called "that early yellow smell" of willows can reliably be savored in March. My own visit to Walden came too early, even for a whiff of willows, but Primack's research has shown that winter conditions matter too. And while I'd missed the T-shirt weather by a day, there was one telling phenomenon that I could easily gauge for myself.

In February 1857, Thoreau estimated the depth of Walden's ice cover at more than two feet (0.6 meter). He tromped across it regularly and described commercial ice cutters making a brisk business with their saws and pikestaffs, hauling off great blue blocks on sleds. When I approached the frozen shore, signs read UNSAFE ICE CONDITIONS, and pictured a stick figure plunging through a hole, arms flailing over its small, round head. Without venturing beyond the shallows, I easily broke through the rime with a stick and plucked out my own tiny ice block. Its thickness measured just under two inches (five centimeters).

Average temperatures around Walden Pond have risen 4.3 degrees Fahrenheit (2.4 degrees Celsius) over the past 160 years, and the typical plant now blooms seven days earlier in springtime. That's a quick shift in biological terms, but if there

FIGURE 3.2. Commercial ice cutters once exported great blocks from Walden Pond, prompting Thoreau to muse that "the sweltering inhabitants of Charleston and New Orleans, of Madras and Bombay and Calcutta, drink at my well." During my February visit, the ice was barely two inches (five centimeters) thick. Photo © Thor Hanson.

weren't more to the story, it would amount to a simple change in expectations—shorter winters, more flowers in April, and perhaps a longer season at Walden's popular swimming beach. Natural systems are rarely that simple, however, and Primack's team quickly spotted another important pattern. The plants blooming earlier, like wood sorrel, tended to be quite common, but the ones that still flowered on the old timetable, including such favorites as fringed orchids and mountain mints, were often hard to find. In fact, many species couldn't be located at all. After years of exhaustive surveys, Primack and his colleagues reached the conclusion that over two hundred plant varieties observed by Thoreau had since disappeared from the Walden vicinity. Some of those losses surely stemmed from development and other human changes to the landscape—more houses, highways, and pollution; fewer wetlands and family farms. But climate change, in the form of warmer, earlier springs, had added an overarching challenge to the mix.

"Flexibility," Primack told me, summarizing one of the key take-home messages from his research. He explained how some species have a built-in capacity to respond to temperature change: greening up and blooming when things get warm, regardless of the date. When the climate was stable, this trait didn't count for much—everyone had adjusted to the same general schedule. But when temperatures started rising, it gave those flexible plants an edge, allowing them to begin growing, flowering, and storing energy for days or weeks ahead of the more conservative species. Many of the slowpokes simply couldn't make up for that lost ground, eventually giving way to their more timely neighbors. In some cases, whole communities suffered. Wildflowers below deciduous trees, for example, once enjoyed weeks of full sunlight before the canopy filled in above them. But it turns out that most of the trees are on the flexible side, leafing out quickly in the warm spring air and casting shade over everything living beneath them. Deprived of that early photosynthetic boost, the flowers now struggle to finish their normal growth and bloom, and some lack the energy to set seed. Increasingly, survival in Thoreau's woods seems dependent on keeping up with the neighbors. As Primack puts it, "Plants that can't leaf out early get out-competed." It's a reminder that climate change alters more than temperature—it affects relationships.

Before leaving Walden Pond, I circled back on the trail to look for the place where Thoreau had planted his bean field. Though the enterprise netted him less than nine dollars in profit, he spent a great deal of time there, using a simple hoe to hand cultivate "seven miles" of tightly spaced rows. (My wife, an avid gardener, doubts that figure.) Heavily forested now, the site gives little sign that it ever produced a commercial crop, however meager. But the arrival of a red-bellied woodpecker showed that there was still food to be had, if you knew where to look. It landed on the gray trunk of an oak snag, hopped upward twice, and then paused, cocking its head and peering around as if to see whether anyone was watching. Apparently satisfied, the bird then plucked a hidden acorn from a deep crack in the bark, hammered it open, and began to feed.

Caching surplus nuts and seeds—and remembering where they are—gives various woodpeckers, squirrels, and other forward-thinking species the rare ability to control their own food supply. When pickings get slim, they can always tuck into their stored reserves. The vast majorities of birds, animals, and insects have no such fallback, however, and must always forage in the moment to survive. This makes it imperative that physically demanding activities like migration and reproduction coincide with periods of abundant food. But as the plant data from Walden make clear, climate change is already playing havoc with timing, and not all species are reacting in the same way. In fact, some organisms aren't even reacting to the same things.

Thoreau once called spring birdsong "Nature's grandest voice," and his knowledge of individual trills and warbles helped him track the annual arrival of migrants from the south. Superficially, the data he gathered on birds look a lot like his plant observations—long rows of handwritten names and dates. But the similarity ends there, because while spring now begins earlier for the average plant, birds continue to show up on the same schedule they used in Thoreau's day. Whether migrating from the tropics or points closer at hand, they take their cues not from temperature but from light, spurred into action by the annual spring increase in day length. And that's something that climate change doesn't alter in the slightest. The resulting divergence sets the stage for what biologists call timing mismatches— nectar-rich flowers blooming before the arrival of hummingbirds, for example, or flocks of hungry swallows missing an expected insect hatch. Whether responding at different rates or to different stimuli, species long accustomed to interacting with one another increasingly find themselves in the right places at the wrong times.

The potential for mismatches is enormous, and it stretches far beyond the woods at Walden Pond. Scientists find similar trends in spring phenology virtually anywhere an old dataset can be scrounged up for comparison. In the American Midwest, that baseline comes from another environmental icon, Aldo Leopold, who noted springtime happenings around his cabin in Wisconsin during the 1930s. Records in the United Kingdom date back at least to 1736, when Norfolk naturalist Robert Marsham began observations for a register titled "Indications of Spring," a sixty-year effort to track the timing of everything from turnip flowers to the first sycamore leaf to the song of "the churn owl" (an old name for the European nightjar). To make matters more complex, spring is not the only season in flux. Primack's team recently turned their attention to autumn, where shifts in the timing of fruit production and leaf fall are upending a whole different suite of relationships, from seed dispersal to the onset of hibernation. Longer summers and shorter winters have consequences too, and all of these phenological changes are more extreme in certain ecosystems. Fall temperatures in Alaska's arctic tundra, for example, recently stood so far above normal, so late in the season, that computers at a climate monitoring station automatically deleted the data as false.

In our own lives, unexpected changes in timing often set off a chain reaction of consequences. When an airline flight is delayed, it can lead to missed connections, late arrivals, canceled appointments, and a rapid reshuffling of plans. If the trip is a casual holiday, we probably have the flexibility to adjust our schedule. But when we are expected to participate in a critical and pre-arranged interaction, like a wedding ceremony or a job interview, the stakes become much higher. Plants and animals face similar challenges in a world of changing phenology, where different reactions among species—compounded across ecosystems—impact a dizzying web of relationships, from competition to predation, pollination, and more. Most of the repercussions have yet to be imagined, let alone studied, but research to date echoes Richard Primack's conclusion about the importance of flexibility. Species that can't adjust quickly face the biggest hurdles, and perhaps none are more at risk than those that rely exclusively upon a single resource or partnership. Few situations fit that description better than the link between specialist pollinators and their host plants, where a change in the timing of either affects the future of both. Such relationships have evolved all over the world, but they often remain obscure and undetected. Luckily, I happened to know of a prime example that played out every spring just a few miles from my house. All I needed to get there was a boat.

With the outboard motor switched off and tilted up, I paddled the last few yards to the beach, floating over a field of submerged boulders. (Borrowing the skiff was contingent upon not denting the propeller.) Not surprisingly, I had the tiny islet all to myself. At little more than an acre in size (0.5 hectare), with a rocky shoreline rising just a few feet above the waves, it wasn't exactly a popular destination. But I'd stopped there on a botanical survey years earlier, and I knew that its tiny meadow contained the plant I was looking for, as well as a healthy population of the only bee known to pollinate it.

Walking up the narrow path from the beach, I passed under junipers and willows bent low from a lifetime fighting the coastal winds of the Pacific Northwest. But it was a calm day now, and sunny—perfect weather for bee watching. And as the trail opened out into the meadow, I could see that my timing was perfect too. There, scattered among the grasses, stood dozens of spikes of the creamy white flowers known as death camas. The name comes from the plant's famed toxicity, a bane to sheepherders as well as the occasional hiker or camper who mistakes its lily-like leaves and fleshy bulb for something edible. Scientists know death camas well, and they blame its potency on zygacine, a potent compound that attacks the heart, the lungs, and, for good measure, the digestive tract. When it came time for an updated Latin name, they gave it what amounts to a taxonomic exclamation point: *Toxicoscordion venenosum* var. *venenosum*, "Poisonous bulb poisonous poisonous"!

Finding a comfortable rock from where I could observe several plants at once, I settled down and began to watch. Below the death camas, a carpet of blue-eyed Marys hugged the earth, and I soon counted three species of bumblebees and what appeared to be a sweat bee foraging happily on the small, bright blossoms. But as the minutes stretched to an hour, not a single insect landed on the camas. This wasn't surprising. Most plants concentrate their chemical defenses in leaves, seeds, roots, or other places likely to be gnawed by hungry attackers, but death camas puts poison everywhere, including into its pollen and nectar. This means that instead of receiving a tasty reward, insects visiting a death camas flower can look forward to seizures, paralysis, and, well, death. This would hardly seem a promising pollination strategy if it weren't for the fact that one local bee species had figured out a solution. By evolving a way to digest and detoxify zygacine, the death camas bee earned itself what amounts to a private dining room, a

FIGURE 3.3. A death camas bee (*Andrena as-tragali*) foraging on death camas (*Toxicoscordion venenosum* var. *venenosum*), one of the many specialist pollinator relationships at risk of unraveling if the timing of the bloom diverges from the flight period of the insect. Photo © Thor Hanson.

prolific source of pollen and nectar that other insects avoid. The plant, in turn, enjoys the dedicated services of a pollinator whose attention rarely wanders. But it all depends upon timing.

I got up to stretch my legs and headed for the southern tip of the little island, where regular doses of salt spray from the waves below kept vegetation to a minimum. The bare soil was perfect nesting habitat for death camas bees, which belong to a large genus of miners that make their homes in tunnels underground. Each female excavates and provisions her own tiny burrow, and her offspring spend the winter there in a form of hibernation, before digging out in the spring to start the whole cycle over again. But even though I got down on my hands and knees, I could find no sign of active holes, or of the turned earth left by emerging bees. In spite of ample death camas in the meadow, it seemed that whatever passed for an alarm clock down in bee world had yet to go off.

Was this the beginning of a mismatch? Certainly the flowers were blooming earlier—observations from a nearby nature reserve showed an average spring advance of two weeks for death camas in a span of only thirty years. (The data were gathered by a series of caretakers living alone in a remote cabin, something that begins to sound like a prerequisite for phenology research.) Ground-nesting bees respond to spring temperatures too, but multiple studies suggest that they do so more slowly than the majority of their floral hosts. Perhaps it's because the soil temperature around burrows takes longer to warm than the air around flower buds, or, like the more conservative plants disappearing from Walden Pond, it may just be how they're wired. Either way, the situation creates a particular challenge for specialists. Every day the death camas bees slumber while their chosen flowers bloom is a missed opportunity. It means less time to forage for pollen and nectar, which, for bees, translates directly into fewer offspring. The situation is fraught for the plants too, leading them to invest precious energy in flowers that may never receive a visit from a pollinator. Wasted time and energy always have repercussions in nature, and it remains to be seen how the challenges of such mismatches will be met—by death camas and its bee, or by the players in thousands of similar scenarios across the globe.

Rather than stare at empty flowers, I decided to spend the rest of my island visit waiting for the bees to emerge. They had to dig out at some point, and it was hard to imagine a nicer day for it. From a winter of unprecedented snowfalls, the weather had whipsawed into record-breaking spring warmth in a matter of weeks—just the sort of extreme swing that has already become a hallmark of modern climate change. Like the bees, I wasn't quite prepared for it, and I found myself shedding sweater and stocking cap as the day heated up. But although I saw plenty of spring activity—from wood ants searching for aphids to a hummingbird decorating her nest with lichen—no death camas bees chose that afternoon to reveal themselves. It would take two more trips to the island before I finally spotted their distinctive rusty-gold forms whizzing between nest holes and flowers, making up for lost time. But sitting there, sweating in that unseasonably warm sunshine, I was reminded of something that Richard Primack had told me, right at the end of our conversation about phenology. Timing mismatches are fascinating and attract a lot of research attention, he said, but there was a much simpler reason why many plants and animals were suffering from climate change: "They just get too hot."

CHAPTER FOUR

The Nth Degree

Climate is what you expect; weather is what you get.

—Anonymous

When I was ten years old, an addition to the family home resulted in my very own bedroom, complete with an electric baseboard heater. I remember the satisfaction of not only having a room to myself, but also the ability to keep it as warm as I liked. I can still picture that old thermostat. The dial showed little tick marks from 40 degrees all the way up to 90 degrees Fahrenheit (4–32 degrees Celsius), but right in the middle of that range—from about 65 to 75 degrees Fahrenheit (18–24 degrees Celsius), all the numbers and marks had been replaced by two words: COMFORT ZONE. From a practical standpoint, this provided a good place to aim for when turning on the heat. But the engineers who designed it had inadvertently done something more, neatly expressing what amounts to a universal rule in biology. For every species, there is a preferred range of conditions within which life proceeds normally, and the precise temperature isn't all that important. Outside that comfort zone, however, every degree counts.

On a warming planet, concern naturally turns to the effects of heat stress and what biologists call the critical thermal maximum. That's the temperature above which an organism simply ceases to function. (There is a critical minimum too, but it tends to get short shrift in an era of rising temperatures.) Not surprisingly, heat tolerance varies widely among species. Running the baseboard full tilt in my childhood bedroom made conditions stuffy but tolerable for *Homo sapiens*, but it would have killed various salamanders, herring, or other creatures that pass their critical maxima well below ninety degrees Fahrenheit (thirty-two degrees Celsius). Part of the difference is innate. Mammals and other "warm-blooded" animals have a greater ability to regulate body temperature than "cold-blooded" groups like amphibians or fish, which rely more on the warmth of their surroundings. But the greater reason why so many different comfort zones exist in nature is far more simple: habitat variety. Life on Earth has adapted to thrive everywhere from hot springs to snow-covered tundra, and from tropical reefs to the subzero brine below Antarctic ice sheets. Given such disparities, one might expect the warming climate to favor creatures already accustomed to life in the heat. But the challenge of extreme temperatures may be hardest for those already living on the edge, and one of the first climate change warning signs came from an iconic inhabitant of deserts.

"They like it hot, but not too hot," Barry Sinervo told me over the phone, describing the lizards he has spent more than three decades studying. In that time, mostly at the University of California, Santa Cruz, Sinervo has made important discoveries about lizard evolution, genetics, mating strategies, and how individuals budget their time and energy. Climate change entered the picture almost by accident, when a conversation with two colleagues revealed that they were all seeing a similar trend. Lizard populations had begun disappearing from their old research sites, particularly

FIGURE 4.1. Every species on the planet lives within a range of preferred temperatures, something thermostat manufacturers neatly encapsulate with the concept and phrase "comfort zone." Minnesota Historical Society.

the hotter, drier ones. Sinervo described this realization as "a gut punch," but it got them all thinking. Was climate change pushing desert lizards beyond their comfort zone? And if so, how?

"I was surprised by how easy it was to predict," he said, and briefly outlined to me the mathematical model they developed. With a few simple inputs about lizards and temperature, the model accurately identified which populations were at risk—not just for the species they were familiar with, but for lizards all over the world. Their publication on the topic has already been cited more than a thousand times, which in science amounts to writing a runaway bestseller. "It's my new calling," he joked. "The Nostradamus of climate change."

Sinervo has a way of describing groundbreaking research with a sort of "anyone-can-do-this" enthusiasm that makes you believe just that. It's no wonder his collaborators and coauthors include everyone from undergraduates to seasoned professionals, in places as far-flung as Chile, China, and the Kalahari Desert. By the end of our conversation he'd roped me in too, with a firm pledge that we would go lizard hunting together the next time I traveled to California. For sheer relatability, it helps that Sinervo focuses on a group of species that many people can see in their own backyards.

Fence lizards and spiny lizards belong to the genus *Sceloporus*, and they rank among the most common reptiles in North

FIGURE 4.2. Fence lizards (*Sceloporus* spp.) and other basking species respond to increasing temperatures by spending more time in the shade, giving up valuable foraging time and sometimes jeopardizing reproduction. Depositphotos/Morphart.

America. Dozens of species inhabit deserts and other warm land-scapes from Mexico north almost as far as Canada. I remember trying—and failing—to catch them as a child. No matter how quickly I lunged, the lizards were faster, darting away from my outstretched hand and under the nearest rock. Experts like Sinervo know that it works better to use a fishing pole, safely capturing your chosen target from a distance by snaring it in a loop of transparent line. Experts also know that lizards rely on rocks for a lot more than evading curious children.

In the language of science, fence lizards and their kin are known as heliotherms—they regulate their body temperature by basking in the sun. This explains why one often sees them stretched out in full view, particularly in the morning or on cooler days when they need a thermal boost to get moving. Too much sun, however, can quickly push them past their thermal maxima, so they never stray far from shady refuges. Moving into and out of direct sunlight is the lizard equivalent of adjusting the thermostat, allowing them to maintain a comfortable and safe body temperature in a range of conditions. As the climate has warmed, lizards have responded by doing what they've always done on hot days—they spend more time in the shade. And that's where the trouble starts.

"It's about what we call hours of restriction," Sinervo said, and explained how his team had uncovered a profound relationship among heat, behavior, and reproduction. Whenever lizards are forced to take refuge from the sun, they give up valuable time that might otherwise have been spent hunting for food. Those lost calories add up, particularly for females during the breeding season. "If it gets too hot, they stop reproducing," he told me. "It's really simple. They just don't have enough energy." Sinervo and his colleagues found this pattern so universal they were able to calculate a precise tipping point: when lizards consistently retreat into the shade for more than 3.85 hours per day, reproduction stops. And it doesn't take a mathematical model to predict the long-term repercussions of that.

For biologists studying climate change, Barry Sinervo's research is a reminder that temperatures don't have to be lethal to be consequential. Certainly there are examples of species exceeding their thermal maxima. Prolonged heat waves in Australia, for example, have caused entire colonies of fruit bats to simply drop dead from their roost trees. But it's more common for rising temperatures to impact how organisms budget their time and energy. Variations of the lizard scenario are now showing up everywhere, from African wild dogs spending fewer daylight hours hunting (and raising fewer pups) to tropical ants abandoning overheated foraging routes through the rainforest canopy. Plants are responding too. Though they obviously can't pick up and move into the shade, species as familiar as the common garden tomato divert energy away from reproduction (i.e., making tomatoes) to stabilize and protect the cells in their heat-stressed leaves. There is another aspect of extreme temperatures, however, that may create even larger challenges. It has to do with how heat—and coping with it—impact the prevalence of disease. And one of the key insights into that relationship came about in an unusual way, when someone turned the wrong valve on a tank full of starfish.

"They were horrified," Drew Harvell said, recalling the reaction of the maintenance crew that accidentally overheated her starfish. "But I told them not to beat themselves up about it. We learned a lot!"

The mishap occurred at a marine biology station not far from my island home in Washington State. It was the spring of 2014 and Harvell, a professor from Cornell University in New York, was coordinating the scientific response to an unfolding crisis. All up and down the west coast of North America, millions of starfish from at least twenty species were dying, their bodies mysteriously twisting and collapsing as if dissolved from within. By the time Harvell saw the evidence for herself on a beach in Seattle (where she'd driven straight from the airport), she already suspected a disease outbreak. Nothing else could have affected so many different varieties over such a wide geographic area. And it was happening fast—entire populations melting away in a matter of months. "I knew we were in deep, deep trouble," she recalled. Her team soon identified a virus (or at least something virus-sized) as the main suspect. Which brings us to the starfish in the tank.

"They were all *Pycnopodia*," she told me, using the genus name for the sunflower sea star, a colorful, many-armed species that can grow as large as a pizza platter and weigh more than eleven pounds (five kilograms). It seemed to be a particularly susceptible species, which made it a good subject for testing potential pathogens. But to identify the cause of a disease you need to start with specimens that don't already have it. "We thought they were healthy," Harvell said, and explained how the stars in the holding tank had all been gathered from locations that still seemed pristine. But before this particular group of *Pycnopodia* could be added to her carefully planned experiment, the maintenance staff inadvertently started another one by shutting off the flow of cool seawater to the tank. Several hours passed before anyone noticed, and in that time the water temperature shot up beyond the starfish comfort zone. At first it didn't seem like a big deal; conditions quickly returned to normal once the water flow was restored. But within days every specimen in the tank withered and died from the same affliction killing them in the wild. The implication was clear—overheating the *Pycnopodia* in warm water had activated the disease, triggering the symptoms of something they were already carrying.

"It's a double whammy," Harvell explained, noting how heat stress weakens a host's immune system while at the same time making its pathogens more prolific. She'd seen it happen before, many times. In thirty years of studying marine disease outbreaks, Harvell had watched rising water temperatures exacerbate maladies in everything from lobsters to abalone. A career that began with the minutiae of disease resistance had broadened into something with global implications—because warming oceans stress the plants and animals that live there, and stressed organisms get sick. That potent connection has given Harvell's expertise a growing relevance she might never have wished for.

My conversation with Drew Harvell took place on a lovely deck, surrounded by flowers, with a big yellow dog asleep at our feet. We met at the home she and her oceanographer husband retreat to whenever they aren't at Cornell teaching, or traveling the world for various research projects and conferences. By happy coincidence, that house lies just a mile or two down the road from where I live, which on a rural island makes us practically neighbors. (Our meeting was the only one conducted for this book that I could reach on a bicycle, and where the discussion veered into complaints about foxes, raccoons, and other neighborhood threats to backyard chicken flocks.) Trim and fit, with bobbed gray hair and an ageless face, Harvell projects a sort of calm thoughtfulness—friendly, but with a lot going on under the surface. She chooses words carefully in conversation, and it's easy to imagine her colleagues and students listening as intently as I did.

"The impacts of climate change are even greater in the ocean than on land," Harvell told me. She'd gained that insight the hard way, by spending a lot of time underwater, taking notes. Many biologists agree with her. Conditions are indeed changing rapidly and unpredictably for life in the oceans, and the synergy between heat and disease has a lot to do with it. Harvell emphasized the decline of tropical corals as another powerful example. Rising water temperatures stress both the polyps themselves and the symbiotic algae they host, resulting in bleached and weakened corals that easily succumb to pathogens. And if those corals die, their loss sends ripple effects through the entire ecosystem. The same thing can be said about starfish. In fact, it was one of the very starfish species now suffering from disease that inspired what has become a fundamental ecological principle—how certain organisms exert an outsized influence on their neighbors.

"It's really come full circle," Harvell admitted, when I asked about the path of her career. "That's part of why I care so much." Her first research position after graduate school had been with Robert Paine, the late ecologist whose iconic starfish experiments had led to the concept and phrase "keystone species." Paine learned that by removing predatory sea stars in the genus *Pisaster* from the intertidal zone, he could change the whole community—shifting it from a mixture that included barnacles, algae, anemones, and limpets to a near monoculture of mussels. Without sea stars there to feed on them, the mussels simply outcompeted everything else and took over. In a sense, warming ocean temperatures and disease are now making that experiment play out on a vast scale.

"The concept that we could lose *the* keystone species . . ." Harvell began, but trailed off. Even for someone accustomed to studying sea creatures in decline, it was a hard thought to put into words. But her expression brightened when I asked if there was any place nearby where I could still see some starfish. She and her students had just completed their annual *Pisaster* survey, and while most populations remained 70 to 90 percent depleted, one site was unexpectedly prolific. Nobody yet knew why, but any sign of recovery or disease resistance was welcome news to everyone. Marine biologists weren't the only people who missed the colorful sight of sea stars clinging to the local shoreline. So I circled the next good low tide on the calendar, and my son and I made plans for a little *Pisaster* expedition of our own.

"I'm getting wet, but it doesn't matter!" Noah crowed, as we scrambled from rock to rock in a pelting downpour. "This is the most fun we've ever had on a beach—twenty-seven, twenty-eight, twenty-nine!" He shouted the additions to our count, pointing to three huge purple starfish tucked into a crevice near the water-line. It pleased me that he was just as thrilled to see them again as I was. But of course sea stars had once ranked among his favorite coastal creatures—what child wouldn't fall in love with a Technicolor, five-armed animal that looked like something drawn by Dr. Seuss? It speaks to the pace of climate-driven change that a nine-year-old can already feel nostalgic for a different era in nature. But our ad hoc survey did indeed seem like a trip into the past as the count continued to grow: "ninety-four, ninety-five, ninety-six!" For whatever reason, this one site was, as Harvell had promised, "just like it used to be."

We were soaked through to the skin by the time we got back to the car, but after finding 408 *Pisaster* sea stars in under an hour, the rain hadn't dampened our spirits one bit. Even better, the starfish had all appeared healthy, their glistening skin firm to the touch and unmarred by lesions. (For those who have never touched a starfish, they feel surprisingly dry and rough, like a lick from a cat's tongue.) There was another reason we didn't mind the rain. Our usually damp region was in the midst of a severe drought, a daily reminder that global warming alters more than temperature alone. Weather extremes of all kinds are on the rise, from dry spells to deluges, severe windstorms, and even cold snaps. Each creates unique challenges for plants and animals that, like excessive heat, push them outside their comfort zone.

FIGURE 4.3. Ochre sea stars (*Pisaster ochraceus*) come in a range of shades, from brown to orange to vibrant purple. The individuals pictured here are healthy, but most populations have yet to recover from a disease outbreak exacerbated by rising ocean temperatures. Photo © Thor Hanson.

Responses vary (this will be explored in later chapters), but there is no denying that some species simply can't adjust to the new conditions. And one of those species was conspicuously absent from our starfish survey.

At such a low tide I had hoped to spot at least a few starfish from deeper waters, namely the *Pycnopodia* sunflower stars involved in Drew Harvell's tank fiasco. They too are keystone predators, helping control populations of kelp-grazing sea urchins. But unlike their intertidal *Pisaster* cousins, the sunflower stars have not shown even a hint of recovery. Scientists now consider them functionally extinct over much of their former range, and the subsequent loss of kelp forests to hungry urchins is held up as an example of how climate impacts on one species can affect whole ecosystems. By any local measure, *Pycnopodia* appeared to be a climate change casualty, but in a comment that was almost an afterthought, Harvell told me something else that added another layer to the story.

When I asked what she would do with a million-dollar research budget, she immediately mentioned setting up shop in Dutch Harbor, Alaska, a remote fishing town located in waters still cold enough to be unaffected by the starfish disease. "*Pycnopodia* are doing great there," she said. It was one of the last healthy populations where studies of disease causation and response could be conducted with confidence. Then she added an observation that I didn't see coming. "In fact, their range is expanding." For sunflower sea stars, it appears that the same temperature trends making life difficult to the south are opening a door northward. The once-frigid waters of the Bering Sea have become less of a barrier, allowing the species to colonize new coastlines throughout the Aleutian Islands and beyond. While this is good news for starfish fans, it raises an obvious question at the heart of another major climate change challenge. We know that *removing* species can have dramatic effects on natural communities—the whole keystone concept proves that. What, then, does it mean when new species are *added*?

CHAPTER FIVE

Strange Bedfellows

*Misery acquaints a man with
strange bed-fellows. I will here shroud till the
dregs of the storm be past.*

—William Shakespeare
The Tempest (1611)

Three orcas surfaced near the shoreline, their dark fins gliding in graceful silhouette against a backdrop of rocks and forest. On a postcard or in a nature film, the moment would have looked serene, but in real life chaos reigned. Dozens of whale-watching boats jockeyed for position behind and around them, drowning out the resonant whoosh of their breathing with engine noise and amplified commentary. The research vessel I was piloting had a permit to get closer. But for this project I kept us right in the middle of the fleet. I was helping out with a study on how boats impacted whale behavior, an important question in such high-traffic conditions. My job was to hold a steady course while spotters equipped with laser rangefinders made a real-time map of the ever-changing scene. Things were proceeding as planned until I noticed something truly remarkable soaring over the pack of boats on stiff, outstretched wings.

"A pelican!" I cried in disbelief. I turned sharply to follow and immediately learned that whale researchers don't take kindly to unplanned avian detours in the middle of their fieldwork. But it was worth enduring a few angry shouts for a glimpse of something so unusual. After all, orcas visited our waters often enough to support a thriving tourism industry. Yet in more than three decades of local birdwatching, I had never before laid eyes on a brown pelican. The bird book mapped its range far to the south, with only occasional strays wandering up the coastline. So I checked it off as a lucky sighting of what birders call a vagrant.

In the years since that incident, demand for my help driving whale research boats has fallen off, but the number of brown pelicans venturing northward has risen higher and higher. Surveys at a roosting site near the mouth of the Columbia River, on the border between Washington and Oregon, rarely found more than one hundred individuals during the 1970s and 1980s. Since 2000, biologists have counted as many as sixteen thousand birds there in a single day, and have witnessed telling signs that the trend is more than a fluke.

"We noticed some of them playing house," Dan Roby told me over the phone, and I could hear a smile in his voice at the memory. "They were picking up sticks and gathering nesting material," he went on, but explained that those first birds were probably juveniles, and that their clumsy attempts didn't amount to much. One pair spent weeks apparently sitting on eggs, but when his team inspected the nest they found a surprise. "It was a fishing lure!" he laughed. "For twenty-eight days they were very assiduously incubating a fishing lure!" By 2013, however, Roby had documented actual eggs laid amid a flurry of courting, mating, and nest building, all on an island more than 900 miles (1,440 kilometers) north of the nearest known breeding colony. "No chicks yet," he cautioned, but if the current trajectory holds, it's probably just a matter of time.

Dan Roby's research put him in an ideal position to witness the arrival of brown pelicans. For over twenty years, he and colleagues from Oregon State University and the US Geological Survey have been monitoring fish-eating birds on the lower Columbia River, part of a larger effort to help manage and protect juvenile salmon headed out to sea. When pelicans started showing up, they simply got counted right alongside the resident cormorants, terns, and gulls, inadvertently documenting a classic example of a range shift. It's a topic we will return to in Chapter Seven as one of the main biological responses to climate change—species following their preferred environmental conditions across the landscape as temperatures warm. Some ranges are expanding, some contracting, and some a combination of both in different areas. For the pelicans, moving north has occurred at a time of overall population growth, and Roby expects them to prosper on the Columbia too. "Food shouldn't be a limiting factor most years," he told me. "They forage out toward the bar in saltwater. Marine schooling fish like anchovies and sardines are abundant out there." Though the birds still can't endure a Pacific Northwest winter, they seem more and more reluctant to head south in the fall, and some models predict their range will extend all the way to Alaska by the end of the century.

The apparent success of brown pelicans in northern waters is only half the story, because range shifts don't just affect the creatures on the move. For local species and habitats, every new arrival amounts to a strange bedfellow, an unknown quantity with the potential to disrupt the status quo. When brown pelicans feed, for example, they plunge headfirst into schools of minnows and fill their massive bills, a habit with obvious relevance for any

FIGURE 5.1. This vintage illustration looks whimsical, but it is based on the very real habit of gulls stealing fish from the overflowing bills of brown pelicans. Competition for food is one of many biological relationships affected when species shift to new places and communities. Depositphotos/Morphart.

sardine or other small fish in the neighborhood. The sudden addition of thousands, or tens of thousands, of such hungry predators also affects the other fish eaters, like gulls and cormorants. How will competition for food change with these new rivals in the mix, particularly during years when prey is less abundant? In some locations, other resources like roosting and nesting sites may also be in short supply. Ornithologists have already raised concerns about how pelicans now dominate several small islands farther north along the Washington coast, places where seabirds like tufted puffins formerly nested. Are the resident birds being crowded out? Scientists find themselves asking such questions more and more frequently, because in an era of climate change, brown pelicans are far from the only animals in motion.

"'Global weirding' seems to be taking over," Roby said, when I asked about other range shifts he'd observed in recent years. White pelicans had also begun showing up at his study sites, while a number of local Caspian terns had pulled up stakes and headed for Alaska. Some of the most extreme "weirding," however, wasn't flying over the waves but swimming and drifting beneath them. Warming ocean temperatures and altered currents have moved so many things from lower latitudes toward the poles, in so many places, that marine biologists are calling the trend tropicalization. Along one stretch of Northern California coastline, recent surveys documented thirty-seven species leaping an average of 215 miles (345 kilometers) north in only four years, including barnacles, sea slugs, snails, crabs, algae, and the bottlenose dolphin. Dozens of additional sightings were deemed so far from home that scientists marked them down (for now at least) as scouts rather than settlers. A two-ton hoodwinker sunfish, for example, wasn't just the first one documented for the state—it was the first recorded in the entire hemisphere.

Calling the mass redistribution of species "weirding" is actually quite appropriate. The word traces to an Old English phrase for fate or destiny, which is precisely what plants and animals are trying to take charge of when they shift their ranges. It also works in the modern sense of strange or bizarre, since it can feel downright peculiar to watch once-familiar natural communities rearrange themselves so quickly. Finally, in the traditional dialect of Scotland, a weird was someone with the ability to see into the future. Biologists like Dan Roby would love to have that knack now, but with thousands of species on the move in ecosystems across the globe, the situation is simply too chaotic for prognostication. Climate models give an indication of where species are going, but it's anyone's guess what will happen when they get there. Some may settle into their new communities with little fanfare, but others have the potential to transform the entire neighborhood. Perhaps nowhere is that drama more apparent than in the thin layer of cells between the bark and the sapwood in pine trees.

In the woods around our house, lodgepole pines get top-heavy. Their branches cluster high up in a tight crown as they grow, making the trunks of older individuals prone to snapping off in strong windstorms. I know this because I look forward to it—not out of any ill will toward pine trees, but because I'm always on the alert for easy sources of firewood. Whenever one falls, I arrive on the scene soon after, axe and saw in hand. But pine logs make up only a small portion of our woodpile, because the local forest is dominated by coastal species like Douglas fir. Farther inland, however, lodgepoles cover vast swathes of western North America, and in those forests the trees have a lot more to worry about than wind and woodpiles. That's because warmer winters have allowed mountain pine beetles to expand their range northward, triggering what has been called the largest insect outbreak in recorded history. It's been going on for years, and I've seen the pictures of hillsides covered with brown and dying timber, but not until working on this chapter did it occur to me to go looking for the beetles close to home. So before starting to write this paragraph, I got a hatchet from the shed and went over to inspect the remains of a pine tree alongside our driveway, figuring that if I gave myself enough time, I might be able to locate signs of beetle damage. It took less than thirty seconds.

Loose bark fell easily from the stump and lower branches, revealing a whole world of beetle trails etched into the wood like looping calligraphy. Some paths split into orderly networks, while others meandered like lost miners. I knew that several species could inhabit the same tree, and that their patterns were often diagnostic. But even with an old US Forest Service pest manual as a guide, I couldn't tell one trail from another. Was there a mountain pine beetle among them? I snapped a few pictures, sent an email, and had an answer within hours.

FIGURE 5.2. To a trained eye, bark beetles leave behind patterns as distinctive as a signature. Staffan Lindgren identified those in my pine tree as an *Ips* engraver (left) and a type of round-headed borer, probably a pine sawyer (right). Look closely at the borer tracks and you can see how they gradually widened as the developing larvae grew in size. Photos © Thor Hanson.

"It is not unlikely that mountain pine beetles may show up there," Staffan Lindgren wrote back, noting that a few had already been located on an island nearby. After spending much of his entomological career on the front lines of the outbreak, Lindgren knows just the kinds of places that the beetles prefer. He also had no trouble identifying the tracks in my photos as the handiwork of others—a type of weevil called an *Ips* engraver, and some sort of round-headed borer. Like most bark beetles, both of these species typically inhabit dead or dying timber, and they had probably invaded our pine after its fall. What really sets mountain pine beetles apart is their ability to infest and overcome perfectly healthy trees. That's why taxonomists assigned their genus the name *Dendroctonus*, a Latin and Greek combination that translates to "tree murderer." When I asked Lindgren what made them so deadly, he replied with his version of a quote from noted Canadian forester Fred Bunnell: "It's not rocket science. It's much more complicated than that."

"The female is the attacking beetle," Lindgren explained over the phone. She bores a hole through the bark and then, if she hasn't mated already, emits a strong pheromone to call in a partner. (Diabolically, she concocts her perfume from defensive chemicals produced by the wounded tree.) Arriving males produce additional tree-derived attractants, drawing more and more beetles of both sexes in a mass attack made much worse by what the beetles are carrying. Lodged within special pockets in their mouthparts are the spores of several distinct fungi that will also invade the tree, spreading blue-tinted rot deep into the sapwood. "The fungal associate is usually assumed to be the killing agent," Lindgren said, and described how the infection gradually chokes off the tree's pathways for moving water and nutrients (as well as resin, its main defense against the beetles). In a final twist, newly hatched beetle larvae then supplement their woody diet by feeding on the fungi, gaining a significant nutritional boost and ensuring that their mouthparts will be filled with fresh spores to carry away to the next unfortunate victim.

Historically, this complex system was largely held in check by weather. Cold snaps in fall and winter killed off the beetles, restricting their range and limiting the size and duration of outbreaks. But rising seasonal temperatures have made those hard freezes increasingly scarce, allowing beetle populations to keep building up, year after year. "When they get to a certain level, they're unstoppable," Lindgren told me, and compared the results to a forest fire burning out of control. "They'll just keep going until they run out of fuel." That stage is known as completion, the point at which the beetles in a given landscape have literally eaten themselves out of house and home. It used to be a rare event, but since the current outbreak started surging northward in the 1990s and early 2000s, it has played out across millions of acres, leaving behind skeletal forests of dead pine trees covering

an area roughly the size of Germany. For researchers like Lindgren, however, there was another curiosity that made this outbreak unusual: wherever the beetles entered territory outside
their historical range, they began to pick up speed.

"They were moving 30 percent faster than the models predicted," Lindgren said, recalling the early years of the outbreak.
Like Dan Roby watching pelicans on the Columbia River, Lindgren was in the right place at the right time to witness the spread
of mountain pine beetles. In 1994, he accepted a position at the
University of Northern British Columbia and moved to the small
city of Prince George, located in a landscape not so different from
his native Sweden. Already a seasoned bark beetle researcher,
Lindgren was also a successful beetle entrepreneur—the Lindgren
funnel trap is a device he invented as a graduate student, and it
remains the gold standard for sampling wild populations. Prince
George is surrounded by lodgepole pine forest, and Lindgren's arrival put him directly in the path of the beetle's northward range
expansion. Here the similarity with the pelican situation unrav-
els, however, because while the birds' impacts may take decades
to understand, the beetles made their presence known immediately. And Lindgren had a pretty good idea why they were ram-
paging so quickly through their new home.

"It was a free-for-all," he said, and described how beetles colonizing new habitat found themselves surrounded by "naïve hosts,"
pine trees with no evolutionary history of resisting attacks. For the
beetles, such trees are easy prey, lacking the defenses that had developed in forests farther south. "Even in a single outbreak, there
is selection going on," Lindgren pointed out, noting how beetles
quickly weeded out the most vulnerable and palatable trees. Over
time, the surviving pines and their descendants had developed
stronger chemicals, copious resin, and other ways to at least slow
down a beetle attack, if not stop it outright. In an elegant proof of

this point, Lindgren and several colleagues simply counted how many offspring the beetles could produce in different landscapes. Where the trees were most naïve, beetle reproduction more than doubled, leading to incredible densities that made the whole system, in Lindgren's words, "go into a sprint."

The concept of naïveté in biology dates back at least to Charles Darwin, who marveled at "the extreme tameness of the birds" in the Galápagos. Living in the absence of terrestrial predators (or curious naturalists), the birds—as well as the iguanas, tortoises, and other creatures he found there—had developed little natural fear of things encountered on land. That lack of evolutionary familiarity made them vulnerable, in that they could be approached for observation, or for the stewpot, with equal ease. The same principle holds true whenever plants or animals encounter other species that are completely novel. They may find themselves without the appropriate defenses, particularly against new predators, competitors, pathogens, and parasites. This does more than account for the speed of beetle outbreaks—it explains why the proliferation of climate-driven range shifts has such potential to reorder natural communities.

For over a decade, mountain pine beetles continued their sprint northward, sweeping past Lindgren's study sites in Prince George and nearly reaching Canada's Yukon Territory, before finally encountering winter temperatures still cold enough to stop them. But climate change hasn't just opened new latitudes to pine beetles, higher *elevations* are also warming, so the outbreak continued by turning east, moving up and over the Rocky Mountains—which had previously been a barrier—and spreading deep into the province of Alberta. Many scientists expect the beetles to carry on all the way across the continent, exploiting naïve hosts (of various pine species) every step along the way. It's an impressive run for a tiny insect that, to paraphrase one noted beetle expert, looks more or less like a mouse turd. And it's only one story among many, because North American pine forests are hardly the only ecosystem struggling to adjust to strange bedfellows. Arctic tundra plants are greeting new herbivores, from moose to moth larvae, and so are Tasmanian kelp beds, in the form of invading sea urchins. Salt marsh grasses on five continents now find themselves competing for space with expanding tropical mangroves, while the soft-bodied seabed creatures around Antarctica will soon be forced to contend with shell-crushing king crabs. So many species are in flux, creating so many novel combinations and communities, that the best guidance for the future may lie in another old adage favored by professional foresters: "Plan for surprise."

When I asked Staffan Lindgren what comes next for lodgepole pines, he rattled off a range of post-outbreak challenges, from severe fires and erosion to altered conditions in the soil and water table. Historically, pine forests would regenerate from the seeds of remnant, beetle-resistant trees, and that may still be happening in some places. But, as Lindgren pointed out, the same warming trend that brought the beetles has also increased the frequency and severity of summer droughts and heat waves—stressful conditions that may render some habitat unsuitable for trees. It also bears mentioning how the effects of the outbreak have rippled outward, pulling and tugging on the proverbial spiderweb of natural connections. Woodpeckers boomed in the immediate aftermath, feeding first on mountain pine beetles and then the various engravers and borers that briefly flourished in all the dead and rotting timber. But life became more difficult for many other species, from deep forest specialists like fishers and goshawks deprived of cover, to squirrels and crossbills that suffered from the sudden lack of pinecones, to woodland caribou confounded in their travels by the impenetrable jumble of fallen tree trunks.

FIGURE 5.3. The western pine elfin (*Callophrys eryphon*). The caterpillars of this butterfly feed exclusively on young pine needles, linking its prospects in a changing climate to those of its host trees. Photo © Alan Schmierer.

As in so many climate change scenarios, it's too soon to say how the mountain pine beetle range shift will play out, and most of its repercussions will never be studied. I've seen no mention, for example, of the consequences for the western pine elfin, a small brown butterfly distinguished—in the right light—by a gorgeous wine-stain wash of purple across its back wings. Elfins are notable as one of a very few creatures able to digest the tough greenery of pine needles. In fact, their caterpillars will eat nothing else. This links them inextricably to the trees, and it brings up another major biological challenge brought on by climate change: What happens when one of life's bare necessities suddenly goes missing?

CHAPTER SIX

The Bare Necessities

Unseen in the background, Fate was quietly slipping lead into the boxing-glove.

—P. G. Wodehouse
"Jeeves and the Old School Chum" (1930)

It was a perfect shot. Bill held the small bird in a perching position and I zoomed in to fill the frame. Snapping such photos added only a few seconds to our well-practiced routine of tagging and measuring specimens from the nearby mist net. Soon this beautiful Sharpe's akalat, with its rusty breast and arching white eyebrows, would be winging its way back into the rainforest. But just as my finger pressed down on the shutter button I felt a sudden downdraft and heard a thwack, a shout, and a rush of wings. When I lowered the camera, the bird was gone and I saw only Bill, clutching his hand and wearing the most startled expression I've ever seen on a human face.

Unbeknownst to us, an African goshawk had been quietly observing our work from a tree branch directly overhead. Raptors aren't known to drool, but it must have been excruciating for a predator of songbirds to watch so many potential meals handled

FIGURE 6.1. The Sharpe's akalat (left) and the African goshawk (right), adversaries and forest dwellers in the Usambara Mountains, Tanzania. Photos © Thor Hanson.

and released right in front of its eyes. Apparently, the sight of that tasty and defenseless akalat, held up like an offering, was the final straw. But something went awry at the last instant and the hawk pulled out of its dive, giving Bill's hand only a glancing blow. The lucky akalat escaped in the confusion and the goshawk flapped off too, suffering little more than wounded pride. But as residents of a mountain forest in a changing world, both birds faced long-term challenges every bit as consequential as the dance between predator and prey.

Jutting up from the plains east of Kilimanjaro, Tanzania's Usambara Mountains rank high among the world's biodiversity hotspots—places exceptionally rich in unique plants and animals. My colleague in the akalat incident was Bill Newmark, a conservation biologist who has been studying Usambara bird populations for more than three decades. (Locals refer to him as *Bwana Ndege*, Swahili for "Mr. Bird.") In that time, he and his field crews have netted, released, and recaptured over thirty thousand individual birds, gaining critical insights into which species survive, and which don't, when a forest is disturbed and divided into fragments. I joined Bill as a master's student to examine one small piece of that puzzle—whether or not rats and other egg eaters were contributing to the decline of birds in smaller patches (they weren't). Throughout my time there, we often came across the stumps of trees harvested for timber, and heard the steady chopping of woodcutters gathering firewood, or farmers clearing land for crops. Our research focused on forest loss and fragmentation because the danger was clear and present, while the effects of climate change still seemed theoretical. We knew the standard prediction that mountain species would follow their preferred conditions higher as temperatures warmed, and that suitable habitat might disappear altogether for those already living on the top. But no one realized how fast that process was playing out until recently, when a young ornithologist took what now seems like an obvious step: he went to the mountains and measured it.

"You have to ask the birds," Ben Freeman explained, tidily summarizing his research philosophy. "Models tell us nothing about what is actually happening in the real world." That itch to get outside apparently defines his interview philosophy as well. We quickly abandoned the spartan office he shares with another postdoctoral fellow at the University of British Columbia, and found a picnic table in a sunny courtyard nearby. (As we talked, I noticed him occasionally glancing over my shoulder at the shrubbery, where two white-crowned sparrows could be seen bringing food to a newly fledged chick.) Standing well over six feet tall (1.8 meters), with a thin frame and distant eyes, Freeman has the unflappable manner of someone accustomed to working in remote locations. But there is nothing remote about his passion for science, and we found ourselves deep into a conversation about the evolution of birdsong before it occurred to either of us that I'd come to see him about something else.

Back on topic, I asked what had spurred his interest in upslope migration, and why he chose to study it in a place as challenging as the wilds of Papua New Guinea. "I got really into Jared Diamond's work," he said, referring to the prominent environmental historian who also happens to be an accomplished tropical ornithologist. Diamond's defining research on the birds of New Guinea included surveys during the 1960s that showed precisely where various species lived on mountainsides. For Freeman, those elevation ranges were more than just a footnote in a bird book. He contacted Diamond and proposed repeating the study—looking for the same birds, on the same mountains, with the same methods—to see if anything had changed.

Months later, and with Diamond's enthusiastic support, Freeman found himself on the slopes of Mount Karimui in the central highlands of New Guinea, setting up mist nets with his wife and frequent collaborator, Alexandra Class Freeman. Also an ornithologist (they met in an Ecuadorian cloud forest), Class Freeman had recently completed her own doctoral dissertation and was now pitching in on what would be the most challenging and rewarding part of her husband's. They hired local field assistants and navigated the complex politics of clan rivalries and obligations. They found a village elder who remembered delivering sweet potatoes to Diamond's camps and helped them relocate the same route up the mountain. They coped with intermittent drinking water, grueling workdays, and conditions that were at times, in Freeman's words, "utterly unsafe." ("It was a good test of a marriage," he observed wryly.) But from a scientific standpoint, everything was falling into place.

"Luckily, the forest was still untouched," he said. With the exception of a one-acre clearing for a cell phone tower, the mountain looked as pristine as it had fifty years earlier, eliminating the possibility that hunting, logging, or other disturbances might have somehow altered the bird community. The only change was a seemingly modest 0.7 degree Fahrenheit (0.39 degree Celsius) rise in average temperature. Was that enough to cause a measurable biological reaction? They wouldn't know until the last bird was netted, identified, and released. "It was really important that we did the study blind with regard to Diamond's findings," Freeman said, explaining that they didn't want to bias their fieldwork with preconceptions. So it must have been a rather tense moment when they finally entered the last piece of data and ran the numbers. Fortunately, the results were unequivocal.

"Almost everything was moving up," Freeman said. For the average bird on Mount Karimui, both the upper and lower limits of its range had risen hundreds of feet in less than five decades. "I almost didn't believe it," he mused, but when they repeated the experiment with another of Diamond's datasets on a different mountain, the trend was even more pronounced. Birds overwhelmingly responded to warmer temperatures by chasing their habitat uphill, just like the models had predicted. If anything, the effect was unexpectedly pronounced, and helped answer a long-standing question about climate change in the tropics. By one school of thought, tropical species should be more resilient than those in temperate latitudes—they were already accustomed to a hot environment. But the Freemans' results suggested the opposite, that the diverse and crowded community of a tropical forest has produced a world of specialists that are highly sensitive to any change in their surroundings. "It's more likely a thousand small pushes than one big one," Freeman told me, when I asked what specific cues the birds were responding to. A whole suite of connections is at play, from changes in the insects and plants they feed upon to altered relationships with competitors and predators. Even disease prevalence may be involved. As a case in point, Freeman himself contracted a life-threatening form of malaria while working on the mountain. It used to occur only in the lowlands, but the mosquitoes that carry it—like birds—are moving uphill.

By any measure, Ben Freeman's doctoral research was a success. It advanced the understanding of an important point of theory, and it produced a highly regarded publication in a top journal. But, like all good science, the project led immediately to more questions. If species were indeed moving higher on mountains, then was the other half of the prediction also true? Would habitat disappear for those already at the top, squeezing them out in what some had dubbed an "escalator to extinction"? The data from Mount Karimui couldn't answer that, because only a handful of birds were restricted to the highest elevations, and all of them were rare. Even Jared Diamond had struggled to find the elusive crested berrypecker, for example, so failing to relocate it wouldn't necessarily have told the Freemans that it was gone. They simply might not have been lucky enough to spot one. Unraveling the escalator-to-extinction question required coming up with another historic dataset from another tropical mountain, one where there were a lot of high-altitude specialists that also happened to be quite common. That's a tall order—the sort of scientific wish list that might take years of searching to satisfy. For Ben Freeman, it took about five minutes.

In biology, doctoral studies conclude with what is called the dissertation defense, a public presentation where students report on their research to a largely congenial audience of professors, fellow graduate students, friends, and other well-wishers. Champagne is often served. Freeman's defense was memorable in part because of its youngest attendee—his newborn child. (Class Freeman kept the baby calm in the back of the room by bouncing up and down on a yoga ball.) But the scientific high point occurred at the end of his remarks, during the traditional congratulatory meeting with his faculty committee. Talk turned to next steps, and when he told them about the dream research scenario he was looking for, his advisor mentioned an old survey from a ridge in the Peruvian Amazon that he'd never gotten around to publishing. Like Mount Karimui, the observations had begun at the base and continued upward through thousands of feet in elevation to the peak. But unlike the mountain in New Guinea, the ridge in Peru boasted sixteen species found only at the top. And in 1985, when the survey had been conducted, eleven of them were common and easy to find. "It was the perfect dataset," Freeman recalled. So, in what must count as one of the quickest postdoctoral turnarounds on record, he came out of his dissertation defense already hard at work planning the next research expedition.

After an exploratory trip to relocate the original survey site, Freeman's work in Peru played out like an echo of the study in New Guinea. (Though with the notable absence of Class Freeman, who opted to stay home with their newborn; and with the addition of multitudes of stingless bees that opted to swarm over him and his colleagues all day long, lapping up their mineral-rich sweat.) Again, the forest had luckily remained untouched by logging or other disturbances, which allowed a direct comparison with the older data. And again, the majority of birds had measurably shifted their ranges upslope. But at the ridgetop, where the rainforest gave way to a dwarf woodland of short, moss-covered trees, Freeman found something new: the extinction escalator

FIGURE 6.2. This elfin forest habitat in Peru looks pristine, but rising temperatures have somehow altered it, driving out high elevation specialist birds like the variable antshrike, the buff-browed foliage-gleaner, the hazel-fronted pygmy-tyrant, and the fulvous-breasted flatbill. Photo © Ben Freeman.

was running at full tilt. Of the high elevation specialists common in 1985, nearly half had disappeared, and several of those that remained were now scarce, restricted to the final stop on the survey, right below the summit. The missing species could still be found on other, higher mountains nearby, but the trajectory and implications of the trend were unmistakable.

"I still don't fully grasp it," Freeman said, raising his hands in a palms-up gesture of bafflement. "It just doesn't seem possible that these remote, pristine places are being affected so strongly." For scientists, skepticism is a natural condition—even when it comes to their own results. But upslope migration has now been documented for everything from moths to tree seedlings, not to mention other bird communities (including those in my old stomping grounds, the Usambara Mountains). And because the shape of a mountain narrows at the top, like a pyramid, then the area occupied by any species moving up also narrows, potentially dwindling to nothing. Part of Freeman's surprise at his data comes from the speed of this process, how scores of range shifts and multiple local extinctions have taken place in only a few decades. But there is also something mysterious about what specific changes cause a species to move, or to disappear altogether. "Why do plants and animals live where they do?" Freeman asked me rhetorically. "That's one of the basic questions Charles Darwin was interested in, and we still don't have a good handle on it."

Since Darwin's time, the concept of habitat has expanded to include far more than simply the landscape setting where a species might be found. Biologists now consider all the environmental variables that make that location suitable, from climate to topography, soils, hydrology, and relationships with the other plants and animals that live there. In some cases, rising temperatures take away an obvious habitat necessity. Melting Arctic sea ice, for example, creates that iconic climate change scenario of stranded polar bears, deprived of their favorite place to roam and hunt for seals. Coral declines are similarly direct, immediately reducing food and cover for a wide range of fish and other reef dwellers. More often, however, it's hard to pinpoint the subtleties that make a warming habitat suddenly untenable. Ben Freeman encountered plenty of short, moss-covered trees along the ridge-top in Peru, but some less visible feature of that elfin forest must have been in flux, and its resident birds were already responding and starting to go missing. High elevation specialists embody the communities most at risk—those with no retreat available, and with no way to replace the lost pieces of their habitat. But species in all sorts of places are beginning to find life's necessities in short supply, and there is one particularly vulnerable group that often gets overlooked by those of us who live on land. I was reminded of it recently while standing at low tide on a patch of rocky island shoreline that has been in my family for generations.

"That's the smartest oyster in the world," my father told me. He pointed with his boot to a mollusk wedged between two low boulders, where it was safe from trampling feet and the hulls of beached skiffs and kayaks. It was also safe from my father, who never met an oyster he didn't want to eat. For years, that oyster had defied him on his own doorstep, growing its rough shell larger and larger with the patient increments of a bricklayer. Its age was part of its wisdom, because old oysters are better at building shells than young ones, assembling calcium carbonate directly from seawater in a sturdy form known as calcite. Larval oysters, on the other hand, spend the first weeks of their lives forming shells from aragonite, a less stable form of the same mineral. That distinction matters for oysters because climate change is making the ocean more acidic, which makes it more difficult for them to build and maintain a shell. So it helps to use the toughest materials available. Unfortunately for marine ecosystems, baby oysters aren't the only organisms that rely on aragonite—so do corals, various plankton, and a wide array of snails and bivalves. And just like birds on mountaintops, if something vital goes missing and their habitat becomes unlivable, they have no place else to go.

To understand ocean acidification, it's useful to remember good old Joseph Priestley, standing over that bubbling vat at his neighborhood brewery. He discovered the principle of carbonation by simply sloshing water from one cup to another, allowing it to pick up gas from the rich air above the fermenting beer. Oceans and lakes interact with the atmosphere in the same way, continuously absorbing carbon dioxide as their surface waters slosh around in the wind. So if the level of carbon dioxide rises in the atmosphere, the level in seawater also goes up, and whenever carbon dioxide mixes with water, some of it forms carbonic acid. That's what causes trouble for shell makers, and it's also why people recommend fizzy drinks like club soda as stain removers. Acids, carbonic or otherwise, are corrosive—they work by disrupting the bonds that hold things together. That's handy if you're trying to remove a blot of mustard from a shirt, but it's problematic for the chemistry of shell formation. The calcium carbonate in shells breaks apart rather easily, something my son and I observed by slowly dissolving a duck egg in a glass of seltzer. Things are even worse for shellfish because acidic water turns carbonate into bicarbonate, a compound that can't be used to build a shell in the first place. So rising acidity creates a real conundrum—shells get more vulnerable at the same time the building blocks to repair or replace them become increasingly scarce.

The first signs of ocean acidity in action came from commercial oyster farms, where larvae were failing to form proper shells. Growers have since learned how to buffer local seawater when acidity levels rise, artificially nursing their wards through the vulnerable aragonite phase. But for shell-making creatures in the wild, no treatment is available, and acidity threatens to take away something as critical to their lives as sea ice is to polar bears— perhaps more so. To study how this dilemma was playing out, marine biologists developed their own research wish list. They needed a common, easy-to-find organism that relied exclusively on a fragile, aragonite shell, and that occurred over a wide range of ocean conditions. The ideal subject turned out to be a group of tiny, free-swimming snails known as sea butterflies.

"As far as zooplankton go, they're quite charismatic," Victoria Peck told me, and she should know. As a member of the British Antarctic Survey, Peck has studied plankton populations everywhere from coastal Greenland to the Weddell Sea. I caught up

FIGURE 6.3. Sea butterflies belong to a large group of swimming snails known as pteropods, from the Greek for "winged foot." They feed upon smaller plankton, and are themselves fed upon by various fishes. The species pictured here, *Limacina helicina*, ranges from upper-latitude oceans to the poles, and measures only a tenth of an inch (2.5 millimeters) in diameter. National Oceanic and Atmospheric Administration.

with her on a Skype call to Quebec City, where she was based for an expedition to the Canadian Arctic. Much of Peck's research involves combing through ocean sediments, reconstructing fossil plankton communities to learn about climate conditions in the past. So it was an interesting change of pace for her to focus on a modern climate indicator, and a bonus that sea butterflies happen to be beautiful, with delicate winged feet and coiled shells that glow like crystal in the light of a microscope.

"It turned out to be fortunate that my background was different," Peck explained, noting her graduate degrees in geology and paleoceanography. "I saw different things." She began her work in the midst of a spate of sea butterfly research, when biologists were documenting the expected shell erosion in lab studies, and had confirmed the process in the wild during a spike in acidity along the west coast of North America. Peck chose to focus on polar regions, where winter sea ice interferes with gas exchange between the water and the atmosphere, causing seasonal peaks in acidity. She, too, found sea butterflies with scarred and pitted shells. But in her experience, that was perfectly normal.

"When you look at fossil gastropods, there are always specimens with signs of damage," Peck told me. Ocean life is risky, and shells get knocked about in all sorts of ways, particularly during failed attacks by predators. When Peck and her colleagues examined the sea butterflies, they saw the same thing, and noticed that erosion from acidity was occurring only where the shells had been previously nicked or scraped. Where shells remained intact they seemed impervious, apparently protected from the acidic water by a thin outer layer called the periostracum. (Snails and many bivalves use this film like scaffolding during shell formation, and it often persists indefinitely unless damaged, providing a varnish-like barrier against the elements.) What's more, the sea butterflies were actively repairing their eroded shells, adding layers of aragonite from the inside to make patches that were up to four times as thick as the original. This, too, reminded Peck of patterns she'd seen in fossils, and it suggested a level of resilience that many of her peers found surprising.

"I wasn't very popular," Peck recalled with a somewhat sheepish laugh. Her results challenged earlier findings, and set off a fiery debate in the world of sea butterfly research. But there was really a lot more in agreement than in dispute. Nobody doubts that ocean acidification is a problem for shell makers. Even when repairs are possible they are costly, diverting energy away from foraging, reproduction, and other vital activities. It also turns out that sea butterflies, like oysters, are particularly vulnerable to acidity during the larval stage; so if current carbon emission trends continue, their hardiness as adults may become immaterial. Finally, it bears mentioning that shell formation is far from the only process impacted by acidity. In an underwater environment, chemistry helps animals regulate everything from smell to navigation, vision, and hearing. Scores of studies have now linked acidification with changes in how fish and other sea creatures perceive the world around them, complicating such basic tasks as finding a mate, finding a meal, finding a home, or avoiding the attention of predators. Tweak the system too far, and many species could find their habitat chemically unlivable from sheer confusion of the senses.

Victoria Peck's work on the resilience of sea butterflies does not take away from the threats posed by ocean acidification, but it does highlight a point that often gets overlooked when we talk about the climate crisis: nature is not defenseless. When conditions change, plants and animals respond. Sometimes those responses fall short, but in other cases it's possible to measure effective adaptation and evolution playing out in real time all around us. The following chapters will explore the array of tools and reactions that species have at their disposal, beginning with an in-depth look at movement, which, like the whole idea of change in nature, is a notion that biologists had a surprisingly hard time coming to grips with.

PART THREE

The Responses

When the winds of change blow, some
build walls, others build windmills.
—Chinese proverb

Like any profession, biology abounds with acronyms. Instead of deoxyribonucleic acid, for example, everyone knows (and prefers) to call the stuff of genetic inheritance DNA. So when it comes to facing the challenges brought on by climate change, experts quickly settled on a new abbreviation—MAD, short for Move, Adapt, or Die. But while that summary captures the starkness of the dilemma, it only hints at the diverse and fascinating ways that species are responding. With examples now playing out all around us, it's clear that movement comes in many forms, that adaptation and even evolution can happen faster than expected, and that for a lucky few, life hardly needs to change at all . . .

CHAPTER SEVEN

Move

Old King Coal was a merry old soul,
"I'll move the world," quoth he.

—Charles Mackay
"Old King Coal" (1846)

G ilbert White was obsessed with swallows. He described their flight patterns and studied their diet. He followed them to nest sites and counted their eggs. He commented on their bathing habits, the shape of their feet, and the fleas that infested their feathers. Swallows captured more of White's meticulous attention than any of the many other creatures described in his 1789 opus, *The Natural History of Selborne*. And one question about them puzzled him above all others. It was the same mystery that had fascinated naturalists as far back as Aristotle and Pliny the Elder: Where did they go in winter?

In the mid-eighteenth century, it was common knowledge that swallows and many other birds appeared and disappeared on a predictable schedule, arriving across Europe in the spring and departing by late autumn. Farmers, hunters, and scholars all understood the timing of this habit, if not the precise logic and method. As a rural English curate of better-than-average means, White had the time and resources to dig deeper. He knew the current thinking on migration, but it was still a debatable theory, competing with persistent and sometimes fanciful notions from antiquity. Swedish taxonomist Carl Linnaeus and one of his students drew from both traditions in their 1757 treatise, *Migrationes Avium*. They noted that geese and ducks were clearly migrants, flying in great V-shaped flocks that pointed like arrows in the direction of travel—north in the spring, and south in the fall. Swallows, on the other hand, most certainly passed the winter submerged in local waterways:

> In the latter part of September they resort in great flights to the lakes and rivers, a single bird first lights upon a reed or bulrush, then a second and a third, until it be bent down with their weight, and sinks into the water with them; they emerge again about the ninth of May, at the commencement of the pleasantest part of the year.

Gilbert White's views were a similar jumble of fact and fiction, and he returned again and again to the question of whether swallows hibernated (underwater, or elsewhere), or whether they journeyed south like waterfowl. He even wrote poetry about it: "Amusive birds!—say where your hid retreat / When the frost rages and the tempests beat." A reluctant traveler, White rarely strayed beyond his beloved Selborne parish, but his perspective on migration—of swallows and other species—benefited from broad correspondence and the expanding reach of empire. He inquired about the movements of ring ouzels in Scotland and Dartmoor, and stone curlews flocking in autumn on the Sussex Downs. He wrote to a sea chaplain to follow up on reports of songbirds resting in the rigging of ships. And when his younger brother was posted to the regiment at Gibraltar, White took full advantage of having a trusted observer at a putative migratory crossroads. He sent reference books, journals, and collecting supplies, and the two corresponded avidly for years. Eventually, though White still dreamed of finding at least a few hidden hibernators in Selborne, he came to embrace his brother's view that "myriads of the swallow kind traverse the Straits (of Gibraltar) from north to south and from south to north according to the season." For good measure, the younger White also reported "vast migrations" of other species crossing to Africa and back, from bee eaters to eagles, vultures, and hoopoes. There were so many, in fact, that Gilbert abandoned his usual precision, and appended his list of confirmed migrants with a telling "etc., etc."

FIGURE 7.1. This illustration shows fishermen netting swallows (and a fish or two) from beneath the ice of a frozen lake. Until the nineteenth century, common wisdom held that swallows did not migrate—they submerged themselves every fall and spent the winter months hibernating underwater. Olaus Magnus, *A Description of the Northern Peoples* (Paris, 1555). Beinecke Rare Book & Manuscript Library, Yale University.

Gilbert White's migratory deliberations capture an idea in transition. Like the other naturalists of his day, White was navigating toward a new understanding of animal movement—where creatures go, how they do it, and why. He famously summarized the driving forces of nature as "love and hunger," and he hinted at another truth when he mused that birds heading south would "enjoy a perpetual summer," and that those heading north again would "retreat before the sun as it advances." Such thoughts resonate now because biologists are once again at a turning point in the study of movement. If White could visit today's world, he would no doubt marvel at the various new techniques for tracking animals—from radio collars and tiny GPS transmitters to reading the chemical traces left in fur, feather, and bone. He would be fascinated by everything that these tools have taught us, not only about migration, but about dispersal, homing, and other habitual comings and goings. Like modern scientists, however, White would then have to contend with the realization that all of these age-old patterns are now in flux, shifting and realigning faster than most observers thought possible. Because in a time of rapid change, it turns out that most species crave familiarity, and a surprising number of them will pull up stakes to go and find it.

"It's mind-blowing," Gretta Pecl told me, sounding amazed in spite of—or perhaps because of—her decades of research on the topic. "We're living through the greatest redistribution of species since the last ice age," she said, and rattled off a few statistics. Over thirty thousand climate-driven range shifts have already been observed and measured, including everything from dragonflies to foxes, whales, plankton, and, yes, Gilbert White's beloved swallows. And that's considered the tip of the iceberg. Scientists estimate that between 25 and 85 percent of *all* species are now in the process of relocating. "Even at the low end," Pecl pointed out, "that's a quarter of all life on Earth."

As a full-time university professor and founder of the Global Marine Hotspots Network, not to mention the Range Extension Database and Mapping Project, and a thriving coalition of researchers called Species on the Move, Gretta Pecl is something of a study in motion herself. I felt lucky she could spare the time for a Skype call, made from a borrowed office in Norway where she'd holed up between a research trip to Finland and a speaking engagement in Sweden, having recently finished coordinating a major international conference in South Africa. All of these activities took place far from her home base at the University of Tasmania. That's where she first got interested in the adaptive movement, or, as she likes to put it, the "shiftiness" of species. It wasn't what she set out to work on, but like so many other biologists now studying climate change, Pecl couldn't ignore what she saw happening in the field.

"I was studying the life history of cephalopods—squid, octopus, and cuttlefish," she said, recalling her doctoral research in the mid-1990s along Tasmania's east coast. By a quirk of regional currents, climate change has been warming the waters there at four times the global average, making it a sort of window into the future. So when Pecl started looking for cephalopods, she also got a sneak peek at the distribution of species in a warmer ocean. "We noticed a lot of new things moving in," she told me. There were snappers, fiddler rays, giant rock barnacles, and long-spined sea urchins—all recent arrivals from the coast of mainland Australia, more than 150 miles (240 kilometers) to the north. At the same time, many local species had begun moving, following the warming trend southward. The situation sparked Pecl's curiosity, which is as voracious as her enthusiasm. She speaks rapidly, but with an inviting warmth and clarity that made our video link

FIGURE 7.2. Long-spined sea urchins (*Centrostephanus rodgersii*) were among the first climate-driven species to catch Gretta Pecl's attention. They have followed warming waters southward from mainland Australia to the east coast of Tasmania, where their appetite for algae has transformed many local kelp forests into rocky "urchin barrens." Photo © John Turnbull.

seem like an easy conversation over coffee. In the research community, that infectious energy has helped make her the sort of person author Malcolm Gladwell calls a connector, someone with an uncanny knack for bringing people together.

"Right from the start I wanted this to be interdisciplinary, and multisystem," Pecl said, which helps explain her globe-spanning list of contacts and collaborators—not just fellow biologists, but economists, lawyers, political scientists, health experts, citizen scientists, and more. When I spoke with her, she'd just spent a week living with a traditional ice-fishing community near the Russian-Finnish border. "The time scale of indigenous knowledge dates back thousands of years," she explained, noting how that perspective can deepen the meaning of biological data. "To them, range shifts feel like an invasion." She went on. "They don't know these species. They have no songs about them, no art." Such insights have given Pecl an unusually broad understanding of what it means when species move, but she also knows the intimate details. So my next question for her was perhaps the most basic one of all. Does it work? Is moving a good strategy for coping with the challenges of climate change?

"It works for those species that can do it and survive," Pecl answered. Then she paused, and I got the impression she was choosing her words carefully. Like many experts I've spoken with, Pecl seemed reluctant to identify particular winners and losers in the climate change struggle. (Ben Freeman made ironic "air quotes" with his fingers whenever the phrase "winners and losers" came up.) Certainly, the sheer number of range shifts under way suggests a general theme—those creatures that can move, do move. And they do it quickly. But while that ability may seem like a big advantage, it's far from a guarantee of success.

"If entire ecosystems were marching along in unison," Pecl said finally, "that probably wouldn't be so bad." Instead, species all respond in their own way—moving at different rates or in different directions, or not moving at all. That scrambles everything, or, as Pecl put it, "throws out the ecological rulebook." It means that even if mobile species are able to relocate to a place with their preferred climate, they still face the significant challenges of settling into a new home. They might have to cope with finding unfamiliar foods, or adjust to new predators, competitors, and diseases, all within a community continuously being upended by the steady flow of arrivals and departures. In other words, it's the "strange bedfellows" conundrum writ large. "At the moment, we've got a pretty good handle on how individual species are moving," Pecl told me, but larger questions remained. "What does it mean for ecosystems?" she asked. "What does it mean when 20 or 30 percent of biodiversity is shifting all at once?" For the first time in our conversation, she sounded a little daunted. Then she laughed. "We're only just getting our shit together on that!"

For scientists like Gretta Pecl, the study of species movement remains as dynamic and full of discovery as it was for Gilbert White in the eighteenth century. But climate change has made it even more consequential, a topic of interest far beyond the drawing rooms of country naturalists. That's because the current wave of range shifts doesn't just change ecosystems, it alters how we interact with them. From farm fields to forests to fishing grounds, the rapid addition and subtraction of species is upending old traditions, and people are eager to know what may be coming next. (As Pecl pointed out to me, even national parks and other protected areas are being affected—"You can't just draw a box around something and expect to keep it like it is.") While there's still a great deal to learn, two consistent trends have already made appearances in this book: First, warming temperatures are driving species toward the poles—north above the equator (like Dan Roby's pelicans and Staffan Lindgren's bark beetles), and south down under (like Gretta Pecl's snappers and sea urchins). The second is that species are moving higher in elevation, ascending mountains, ridgelines, and other topographic gradients (like Ben Freeman's birds). Beyond these general patterns, however, surprising examples are emerging. They serve as reminders that species move for all sorts of reasons, and temperature is not always the driving force. One such trend is occurring all across eastern North America, and it's playing out in a group of organisms that aren't typically known for moving. More often, they are admired—even revered—for their reliable stability.

The nine worlds of Norse mythology include separate realms for fire, fog, people, giants, and a range of gods and immortals, all positioned neatly among the branches and roots of a great tree with an equally great name, Yggdrasil. The assumption, of course, is that Yggdrasil can be trusted to stay put, keeping the various worlds distinct and in order for all eternity. Similarly immobile trees feature widely in legends and stories, from the fig that sheltered a meditating Buddha to the stationary apple that dropped its fruit into the gravity field of Isaac Newton's backyard. Shakespeare famously linked the demise of Macbeth to the impossible notion that "Birnam Wood to high Dunsinane hill shall come against him," but when that fateful day arrived, the moving wood turned out to be ordinary soldiers camouflaged by "leafy screens." Even the distinctly ambulatory Ents of J. R. R. Tolkien's *The Lord of the Rings* weren't actually trees, but treelike creatures charged with protecting and caring for the real things (which presumably needed that protection because they couldn't move). In spite of such widespread assumptions, scientists are now finding that trees respond to climate change with the same kinds of rapid range shifts being measured for things like birds and fish. The key to spotting that motion lies in knowing where and how to look.

On a brisk and breezy autumn day I hiked downhill from the rim of the broad Des Moines River valley in central Iowa, following a wooded hogback that curved and descended past a series of sandstone bluffs. Though best known for its cornfields, the state of Iowa also contains a fair number of trees. It lies just past the transition from eastern hardwood forest to the great prairies of the American Midwest, and strips of rich woodland snake out into the plains along every creek and river bottom. With sunlight brightening the furrowed trunks and branches above, and streaming through a canopy of colorful fall leaves, everything seemed designed to draw my attention upward. But I forced my gaze to the ground. The story I'd come to see had less to do with the tall trees arching overhead than it did with the little ones just getting started. In a forest, the old trees tell you about the past; the young ones tell you about the future.

Branching from the main trail, I hiked into the undergrowth and began an informal survey of saplings. There were broad-leaved sugar maples as high as my shoulder, and slippery elms with leaves that felt like sandpaper. I quickly found basswood, hickories, hop hornbeam, and a variety of oaks. Such diversity seemed a far cry from the soggy coastal woods I was used to, where a handful of conifer species blanketed vast acreage. But these young hardwoods weren't just different from a forest half a continent away; they also differed from the mature trees growing directly overhead. That's not to say I couldn't find adult maples, hickories, and others nearby—the seeds had to come from somewhere. But if I took a census of those adults and compared it with the saplings, the *proportion* of species would be noticeably different—some varieties were becoming more common in the new generation, and some less so, suggesting what climate scientists have known for a long time: the environmental conditions for sprouting and growing are markedly different now than they were decades ago, when the adults got their start. Gather enough such data and it becomes possible to see exactly what so terrified Macbeth: armies of trees marching rapidly across the landscape.

By the time I stopped for lunch, my tally of saplings included seventy-five individuals from sixteen different species, all of them deceptively still. To be sure of which ones were moving, and how fast, I'd brought along a research paper by Purdue University professor Songlin Fei. Unlike Gretta Pecl's or Gilbert White's, Fei's interest in movement didn't arise from observations made in the field. His specialty is computational ecology, the study of patterns in nature through mathematical modeling, computer simulations, and the careful analysis of complex data. "I had been studying large-scale forest ecology questions for over a decade," he told me in an email, and described looking for climate change impacts as the next logical step. But where most studies focused on predictions, his team wanted to show what was already happening. "People often have a hard time connecting to models of projected risks," he explained. "These are scenarios for the future, and models often have large uncertainties. We wanted to show how climate change has already impacted forest ecosystems, using a long-term, large-scale dataset." Fortunately for Fei, just such a dataset already existed, and it was available for free online from the United States Forest Service.

The Forest Inventory and Analysis Program describes itself as "the Nation's Tree Census." It compiles annual surveys that amount to a formalized version of my walk through the Iowa woods—counting, measuring, and identifying saplings, as well as adult trees, across a vast network of forest plots scattered around the country. Fei's team downloaded all the data stretching from the Midwestern states east to the Atlantic coast, dating back as far as 1980. (To put that amount of information into perspective, I tried to download the data for Iowa alone. It included 71,025 rows of numbers divided into 182 columns—the spreadsheet program on my laptop couldn't even open the file.) Operating at such a vast scale allowed Fei's team to find what he calls the geographic center for each species. It's like a midpoint for the whole range, the physical location where that species reaches its peak abundance in eastern North America. The math is complicated, but the concept is familiar. When baseball fans go to a game hoping to catch a fly ball, for instance, they make a very similar calculation in choosing their seats. Batters send drives and pop-ups into the stands all over the place, but the geography of where those balls land has a clear center, a hot zone where the likelihood of making a catch is highest. It's easy to imagine those baseballs behaving differently under different conditions—a strong wind, for example, would push them all in one direction, and so would a batting lineup full of left-handed hitters. Tracking the geographic center is a perfect way to measure such movements because it captures the behavior of the population as a whole, telling fans just where to sit and hold up their empty gloves. For trees, Songlin Fei and his colleagues expected to see movement in response to climate change, and they were right. Nearly 75 percent of the eighty-six species they examined had shifted significantly between 1980 and 2015. The surprise came in where those trees were going.

"We expected to see a northward movement as reported by other studies," Fei explained, and indeed the geographic centers for many trees were headed in that direction. But they found an even larger number moving *west*, prompting two immediate reactions: (1) a full reexamination of their analysis to make sure it

FIGURE 7.3. The Forest Inventory and Analysis Program began measuring tree populations in some parts of the United States as early as 1928. Though conceived as an aid for planning timber harvests, the project has amassed an incredible dataset for studying how trees respond to climate change. United States Forest Service.

was sound (it was); and (2) a thorough search for the cause. They found their answer in patterns of rainfall and drought. "Moisture plays a critical role," Fei wrote, and described how tree populations were moving farthest and fastest in places like Iowa and other Midwestern states, where total annual precipitation had risen by more than half an inch (150 millimeters). This echoes a study from California, where plants have been moving downslope rather than up, again following changes in rainfall rather than temperature. It's a reminder that species respond to a range of variables, and that climate change affects a lot more than just how hot it gets on a particular day. Warmer air in the atmosphere moves differently, with the potential to hold more moisture, altering the timing and intensity of everything from rain and snow to droughts, storms, and wind events. Any or all of these can play a role in determining the suitability of a given location for a particular species. In the case of Fei's trees, having enough available moisture was a stronger pull than the attraction of warmer weather. But while this invites a closer look at *why* trees move, it's also important to ask *how* they manage to do it.

As I hiked back toward my car, I was brought up short by the harsh nasal racket of blue jays, calling out from somewhere deep in the undergrowth. I paused to look, then noticed how the canopy in this part of the forest was almost entirely dominated by oak trees. That seemed fitting, since the interaction between jays and oaks played a key role in working out the dynamics of long-distance movement in trees. It began in 1899, when British geologist and botanist Clement Reid noted a confounding impossibility. Twenty thousand years earlier, ice age glaciers had scoured the British Isles to bare rock, yet now they were well covered in trees. To Reid, the idea of forests returning so quickly simply didn't add up. "The oak, to gain its present most northerly position in North Britain . . . probably had to travel fully six hundred miles, and this without external aid would take something like a million years." He made that calculation based on the short distance an acorn might fall in a windstorm, or if it were carried off by a squirrel. Moving in such tiny hops from southern Europe—where oaks and other trees had persisted throughout the last ice age—would indeed have been a time-consuming journey. The same problem applied to any tree with "large, soft seeds which cannot be carried in fur or feathers, and that are killed by digestion," a designation that included other common varieties like beech and elm. Botanists soon found additional examples of unexpectedly rapid tree dispersal, and they dubbed the phenomenon Reid's Paradox. Long-distance transport by birds appeared to be the only explanation, something Reid himself mused about when he came upon a flock of rooks feeding on acorns in an open field, far from any adult oaks. But it would be nearly a century before observations caught up with theory and put the matter finally to rest.

FIGURE 7.4. Blue jays contribute to the rapid, climate-driven migration of oak trees by carrying acorns long distances into new territory and burying them for later retrieval. Inevitably, some are forgotten and go on to sprout. Photo © Melissa McCarthy.

During the 1980s, improved field techniques finally allowed ornithologists to put some numbers on the great passion that blue jays feel toward acorns. A flock of only fifty birds managed to remove more than 150,000 acorns from a grove of pin oaks in a single season, carrying them away to store for the winter in caches tucked under leaves or buried in the soil. Other studies tracked jays regularly toting acorns from parent trees to sites 2.5 miles (4 kilometers) away and caching them in habitats perfect for germination. What's more, the birds chose only the healthiest and most viable seeds, judging the quality of each acorn by gauging its weight and tapping it with their bill, just like shoppers selecting the ripest melons. These discoveries confirmed that blue jays could indeed have propelled oak forests across the postglacial landscapes of North America. Fossil and pollen records showed the trees advancing at a rate of 2.2 miles (3.5 kilometers) per decade—hard going for a squirrel, but an easy feat for fast-flying jays. The model worked for Britain too—where rooks and European jays played a similar role—and it was easily adapted to explain the rapid spread of other bird-dispersed plant species. Instead of a slow diffusion, botanists began to see plant migration as a dynamic process of long leaps and backfilling, where chance events like storms could complicate things further by dispersing windborne seeds at even greater intervals. Though rooted in an effort to explain past events, these are powerful ideas in the era of modern climate change. They make Songlin Fei's results even more striking, since trees appear to be moving a lot faster now than they were in the wake of ancient glaciers.

Red oaks and white oaks like those I saw in Iowa are chugging along at more than ten miles (seventeen kilometers) per decade, nearly three times their estimated post–ice age rate. Hop hornbeams are even faster, at twenty-one miles (thirty-four kilometers) per decade, which is nothing compared to the honey locust, whose geographic center is whizzing along westward at forty miles (sixty-four kilometers) per decade! Fei's team parsed the numbers further and found that saplings were responding the quickest, which makes sense because germination and establishment are particularly vulnerable times. The data showed young trees surging into new territory wherever conditions were improving, and diminishing in the places where conditions were getting worse. Adult trees were less sensitive, but they followed the same general pattern in rates of survival and persistence. Perhaps the best way to put Fei's results into perspective, however, comes from a similar study on a far more intuitively mobile group: birds. Using data from the National Audubon Society's annual Christmas Bird Count, ornithologists have shown that winter ranges for North American birds are also shifting in response to climate change, but at the comparatively stately pace of just over a half mile (one kilometer) every decade.

The fact that trees can sometimes relocate faster than birds is a reminder that movement in nature doesn't have to be obvious to be significant. Keeping up with climate change isn't always about flying, running, or swimming to a new location. It can be as subtle as improved germination where the soil remains moist, or slightly higher winter survival where temperatures stay mild. Nor does movement happen in isolation—it's not the only way that plants and animals respond. "People say 'move or adapt' as if it's one or the other," Gretta Pecl told me, "but they're not mutually exclusive. Species are moving and adapting at the same time." As we'll see in the next chapter, adaptation can determine where species go and why, or whether they need to bother changing locations in the first place.

CHAPTER EIGHT

Adapt

*Nature's verdict is "Adapt or die." She cherishes her
darlings, the adapted; but the unadapted she disinherits.*

—Thomas Nixon Carver
"The Principle of Self-Centered Appreciation
Commonly Called Self-Interest" (1915)

W hen big bears charge, little bears run. It's the sort of natural law that doesn't usually require much analysis. After
all, a full-grown North American brown bear, also known as a
grizzly, can weigh more than 1,200 pounds (500 kilograms) and
reach speeds exceeding thirty miles per hour (forty-eight kilometers per hour). Large males regularly chase smaller competitors
away from prime feeding sites, which is precisely what I had just
observed on the wide delta of a salmon-laden stream. Things di-
verged from normal, however, when the small bear, followed by
its bulky pursuer, turned and began to sprint directly toward the
people I was there to chaperone.

As a ranger for the United States Forest Service, my job at
Alaska's Pack Creek Bear Viewing Area involved watching over
small groups of tourists that arrived on floatplanes from nearby
Juneau. Today's bunch seemed thrilled by the approaching bears,

grinning and snapping photos as if there were more than empty mudflats between them and the two agitated bruins. (Sometimes people were less trusting and had to be reminded not to flee into the woods.) I was also responsible for collecting data as part of a long-term study on how the bears responded to tourism. This encounter would be logged as a "contingent interaction," an example of typical aggression between bears spilling over to include their human observers. We'd seen it happen before. Some of the younger animals had learned to run toward the viewing area as a form of self-defense, knowing that the dominant males were skittish around people and reluctant to get too close. Sure enough, the big bear soon veered off and returned to the creek, leaving the young one to safely saunter past us, huffing a bit, but unharmed. It was a novel and effective strategy, what biologists would call adaptive behavior—adjusting one's habits to take best advantage of a new situation. We marked it down as an interesting footnote, never suspecting that climate change would soon trigger a much more significant behavioral shift, one at the heart of the relationship that connects bears to salmon streams in the first place.

When you buy a salmon fillet at the grocery, or order one in a restaurant, you're getting part of the powerful swimming muscles that line either side of the backbone and ribs, from just behind the gill cover all the way down to the tail. In a fish-eating family like mine, people jockey for pieces close to the belly, where the fat content can be five times as high as the rest of the body. Bears know this trick too, and they often high-grade their catch by eating only the belly meat, or other choice bits like the brain and the roe. I've also watched bears pin the tail of a fish to the ground with their front paws, and then use their teeth to neatly peel off strips of fatty skin. For people, eating the richest flesh is a matter of taste, but for bears it boils down to nutrition, and an urgent need to put on body weight.

Brown bears are omnivores, adapted to a surprisingly wide variety of foraging strategies. In addition to fishing, coastal individuals are known to graze on grasses and sedges, pick fruit, and even dig for clams, while interior bears snack on everything from moth larvae to rose hips. Fed buffet style in captivity, bears consistently choose a mixed diet dominated by carbohydrates or fats, where protein makes up around 17 percent of the available energy. That ratio maximizes weight gain, an important consideration for animals that spend half the year torpid in their winter dens, living on whatever reserves they've stored up in their muscles and body fat. When bears eat salmon all day they get plenty of calories, but even if they focus on the richest morsels it's still a diet way too high in protein—70 or even 80 percent. Technical papers describe this situation as "suboptimal," but one researcher I spoke with put things more vividly when he said that it gives the bears "terrible diarrhea, and all the rest." Still, the sheer abundance of salmon overwhelms any nutritional shortcomings, and eating fish has long been considered an essential bear habit. Even among experts, it's practically axiomatic that bears love salmon. But on Kodiak Island, 700 miles (1,125 kilometers) west from where I worked at Pack Creek, climate change recently put that assumption to the test, and a group of field biologists had the good fortune to be on hand to watch.

"It happened right in front of our eyes," Will Deacy explained. "The bears all just picked up and left the streams." I had called him to ask about the summer of 2014, when his doctoral research took an adaptive shift of its own. As a wildlife student, Deacy had done, in his words, "the typical thing," working a series of short-term stints on species like stick bugs and tortoises before discovering his particular passion for the bears of Kodiak Island. But he hadn't intended to focus his dissertation on the impacts of climate change. (Among graduate students, he confided: "That's

FIGURE 8.1. Historically, the relationship between bears and salmon has resulted in a lot of well-chewed carcasses like this one. That has begun to change on Alaska's Kodiak Island, however, where bears are abandoning their fishing streams to feed on early ripening elderberries. Photo © Thor Hanson.

almost a cliché now.") Instead, he set out to document how bears prolong their fishing season by moving from watershed to watershed, tracking differences in the timing of the various salmon runs. Everything was going as planned. He had darted nearly forty wild bears and fitted them with GPS collars. He had set up time-lapse cameras to monitor salmon numbers on four critical streams.

But then, just as the salmon began reaching peak abundance, his study subjects suddenly stopped fishing and exited the stage.

"We were lucky we had the tools in place to document everything," Deacy recalled. Because they were already counting salmon, they knew that the bears hadn't left for want of food. And because they had equipped the bears with collars, they could simply follow along behind and see exactly what the animals were up to. Without exception, the missing bruins had abandoned their fishing streams and climbed into the hills with one thing on their minds: berry season. To be clear, there is nothing unusual about bears eating berries. Blueberries, crowberries, and other tiny fruits are rich in carbohydrates and have always been an important source of late-season calories. But in 2014, and again the following year, warm weather triggered an early harvest of a berry that bears apparently prefer above all else, even salmon.

"Elderberries are weird," Deacy said. At first I thought he was referring to their odor, a feature made famous by the Monty Python comedy troupe, who coined a peculiar insult about "smelling of elderberries." Although they do have a slightly offputting mustiness, and are reputed to induce nausea in people when eaten raw, the red elderberries of coastal Alaska boast a *nutritional* weirdness that makes them perfect for bears. Where most berries hardly contain enough protein to measure, in red elderberries that figure stands near 13 percent—remarkably close to the 17 percent preferred by bears in feeding trials. And since the rest of an elderberry's calories come in the form of carbohydrates, bears can fatten up on them more quickly than just about anything else in their diet. Historically, this near-perfect food had been hiding from biologists in plain sight, mixed in with the other berries and fruits that coastal bears turn to in autumn, when the salmon runs begin to wane. Deacy's team only made their unexpected discovery because climate change has now altered the playing field. For elderberries, early spring warming and hotter summers have forced a change in phenology, advancing their flowering and fruiting schedule by more than two weeks. Ripe berries are increasingly available right in the middle of salmon season, forcing bears to make a choice: stay on the old fishing schedule and miss out on their favorite fruit, or change their behavior to keep up with the times.

"It's the right move for the bears," Deacy told me. Provided they continue to find enough food later in the season, he saw no reason why switching to elderberries at the expense of salmon would do them any harm. In fact, he speculated that Kodiak's brown bears, already famous for their size, might even grow larger on their adjusted diet. "The bigger story is how this will affect other species," he said, underscoring one of the central themes in climate change biology—how small shifts in one relationship can have cascading effects on others. By eating fewer salmon, the bears will drag fewer carcasses onto stream banks and into the surrounding woods, reducing food for a variety of scavengers and limiting an important flow of energy from the ocean into land-based systems. (Rotting salmon fertilize the soil, accelerating plant growth and contributing nitrogen, phosphorous, and other nutrients that move throughout the entire food web, from grazers to their predators and beyond. Even songbirds and spiders near salmon streams have measurable levels of salmon-derived nutrients in their bodies.) Deacy expects the vegetation and biodiversity alongside Kodiak's streams and rivers to look quite different in fifty or a hundred years, a change driven in large part by the palate (and adaptability) of bears.

Before our conversation ended, Will Deacy added a caveat. "I want to draw a line between generalist omnivores like bears and other species," he said, and explained how a broad diet and high mobility made brown bears particularly responsive to changes in their environment. Switching from salmon to berries was as easy as walking uphill, and they could do it on whatever date the fruit happened to ripen. Or, in the event of a poor berry harvest, it was just as simple for the bears to go back down to the stream and start fishing again. Sedentary species, or specialized feeders with limited diets, lack such options, and so are far more likely to struggle with the effects of rapid warming. "Climate change favors omnivores and generalists," he stressed, echoing another basic lesson for unsettled times: flexibility matters. In biology, it's a concept important enough to merit its own vocabulary, starting with a word that, ironically, people usually associate with products made from fossil fuels.

Quality Comics debuted the character Plastic Man in 1941, soon after the invention of polyester and Teflon, and at a moment in history when things like Plexiglas windows and nylon stockings were still expensive novelties. The publisher hoped to tap into a wave of popular excitement about these new synthetics, and early covers featured the dramatic slogan "He Stretches . . . He Bends . . . He's PLASTIC MAN!" The red-suited crime-fighter's shape-shifting powers proved to be as handy and enduring as his namesake, earning him hundreds of appearances in print and film, as well as that ultimate superhero honor, membership in the elite squad known as the Justice League. (The same cannot be said for his early sidekick, Woozy Winks, who dressed in green and drew on magical powers and protections bestowed on him by Mother Nature.) Biologists should feel no surprise at the success of Plastic Man—the advantages of flexibility in nature have long been clear. Since at least the 1850s, experts have used the word *plasticity* to describe what can sometimes amount to a superpower for plants and animals: the ability to stretch and bend their habits, and even their bodies, in response to environmental change.

In a broad sense, plasticity refers to adaptation in real time: the various adjustments that an individual can make within a single lifespan. (Adaptation can also be evolutionary, playing out through genetic change over generations; we'll get to that in the next chapter.) When bears alter their diet, that behavioral shift is a form of plasticity. But plastic changes can also be physical, something we all know intuitively from our own experiences as we adjust to changes in the weather. Having grown up in the cool and damp Pacific Northwest, I remember the shock of moving to Southern California for college. At the time, the region was in the midst of a late summer heat wave, with temperatures well above one hundred degrees Fahrenheit (thirty-eight degrees Celsius). "It'll thin your blood, living down there," my father warned me. Of course that's not true—California blood is no more watery than any other. But he was right to suspect that my body would measurably adapt to its new environment. Subtle decreases in heart rate and oxygen consumption, as well as sweat dilution and heightened blood flow to the skin, are all part of the body's acclimatization process to hotter weather. Within a few weeks these unconscious physical adjustments, combined with the behavioral adaptation of wearing shorts and a T-shirt every day, made life in the California sunshine feel perfectly normal.

Coping with weather is of course short-term and reversible, but plasticity can also produce permanent alterations, particularly when it plays out early in life. Growth potential is perhaps the best-known example. For many species, including our own, adult body size is based in part on cues received during the first stages of development. Environmental stressors like poor nutrition can set a trajectory for limited future growth that remains fixed, even if conditions improve later on. This reaction is considered adaptive because it acts like an early warning system for the growing body, signaling the presence of harsh (or at best, unpredictable) surroundings that may lack sufficient food and other resources necessary to support a larger size. Historians and biologists believe this explains the close link between human height and living conditions. People in developed countries have grown taller in recent centuries not through genetic changes, but through plasticity, as improved nutrition for mothers and children (in other words, a change in their environment) triggered inherent pathways for larger size.

There is no doubt that plasticity helps plants and animals roll with the punches on a changing planet, but it's far from evenly distributed. Some species have a lot of plasticity, with a wide range of potential physical and behavioral responses already built into their genetic code. The seed of a common dandelion, for example, will grow into quite a different plant under different conditions, blooming at ground level to avoid a mower blade, for example, or shooting up nearly three feet (0.9 meter) in an open field. Dandelion leaves can be toothed and full of bitter latex when the plant is growing in a dry, gravelly roadside, or tender enough to serve as salad greens when found in a well-watered lawn a few feet away. Dandelions can flower during any month, live for a year or a decade, and produce thousands of seeds without the need for pollination. Such traits make them the bane of weed-weary landscapers, but in the context of climate change their plasticity is like an insurance policy, a hedge against an unpredictable future. The closely related California dandelion, in contrast, exhibits very little plasticity. It blooms exclusively in early summer, requires bees for cross-pollination, and grows only on the fringes of wet, subalpine meadows. Though the two species look nearly identical, differences in plasticity help make one ubiquitous and resilient, while making the other critically endangered, and restricted to a handful of precarious locations in the rapidly warming San Bernardino Mountains.

FIGURE 8.2. Plasticity lies as close to hand as the nearest bunch of common dandelions (*Taraxacum officinale*). It took me only a few minutes to find this range of shapes, sizes, and even colors in the full-size leaves of mature individuals (blooming or in bud) growing near our home. The differences represent the plant's inherent response to different growing conditions, from driveway to footpath, open field, and shady lawn. Photo © Thor Hanson.

While the benefits of plasticity can be substantial, its effects are often rather modest to the naked eye. A dandelion with toothier leaves still looks like a dandelion, and a bear munching on elderberries remains quite obviously a bear. The same is true for countless other adjustments now under way in response to climate change—the species involved make modifications, but they persist in very recognizable roles within their respective communities. In certain cases, however, plasticity reaches surprising extremes, showcasing just what some creatures are capable of when their environment changes. The Humboldt squid, for example, virtually disappeared from traditional fishing grounds in Mexico's

Gulf of California after the water warmed substantially in 2009 and 2010. Or so everybody thought, until surveys found the squid not only still present but more abundant than ever. Instead of departing, they had responded to heat stress by adopting a radically different life strategy. They matured and reproduced in half the time, ate different foods, and lived only half as long. As a result, their new adult bodies were a fraction of their former size—often too small to bite on the lures previously used to catch them. Fishermen had been dismissing the few they could hook as juveniles, or another species altogether.

When body size comes into play, extreme plasticity is easiest to spot in species that grow quickly, like the Humboldt squid. If one or even two generations mature in a single year, different body shapes and dimensions quickly become apparent. But plasticity can also affect behavior in transformational ways, making even well-known species seem utterly different. In mid-2016, mass coral bleaching across the western Pacific caused one of the most prominent and aggressive groups of reef fish to more or less trade in their old personalities overnight, and just like Will Deacy with his brown bears, a group of marine biologists found themselves in the right place at the right time to see it happen.

Sally Keith has watched a lot of butterfly fish. For a marine biologist interested in competition, they make an ideal subject—feisty, territorial, and bright enough to stand out, even in the colorful melee of a tropical reef. Many species feed on coral, vigorously defending their small patches against all comers. Hardly a minute passes without a fight and a chase, which helps to keep the data collection lively. That's not an idle concern if you're going to be spending hundreds of hours in the water with a clipboard, staring at fish. Keith and her colleagues were doing just that in 2016, spread across more than a dozen reefs from Indonesia west through the Philippines and north as far as Japan. The idea behind working at such a large scale was to see whether competition increased near species boundaries. If different types of fish fought more where their ranges overlapped, that interaction might help determine where and why one variety ended and another one began. It's an interesting question, but one that would have to go unanswered, because halfway through the project a marine heat wave sent water temperatures soaring, and things took an unexpected turn.

Reef-building corals may be nature's best-known example of a "mutualism," the sort of relationship where two independent organisms both gain something from a shared interaction. Individual coral polyps are tiny animals distantly related to jellyfish. (Many polyps group together to form what we think of as a single coral.) Like jellyfish, coral polyps also have tentacles and some manage to feed themselves by catching plankton or small fish. But the majority of their diet comes from the even tinier creatures that live within them, a group of single-celled swimmers called dinoflagellates (also symbionts, symbiotic algae, or, for even more Scrabble points, zooxanthellae). Like plants, these little green and red plankton use photosynthesis to make sugars from sunlight. They share that energy with their hosts, and in return they gain a safe home in a sunny location.

Relying so heavily upon sunlight explains why reef corals thrive best in the shallows, skirting lagoons, ringing atolls, and sticking close to shorelines. But it turns out that the temperature of the water is every bit as important as its depth. If things get too hot, the corals and their dinoflagellates become like roommates arguing in a stuffy apartment. Eventually, they part ways, a process known as bleaching because it leaves the corals ghostly pale in the absence of their colorful partners. Bleached corals can survive temporarily, and sometimes they even reabsorb dinoflagellates if the water cools off in time. But prolonged heat waves lead to coral disease and starvation, and climate change is making such deadly episodes more and more common. The whole reef system suffers in the aftermath, and many studies have documented dramatic die-offs and steep declines in the diversity of fishes and other reef dwellers. But Keith's team became one of the first to witness and measure species quickly adapting to the new conditions. And it made their fieldwork extremely tedious.

"After the bleaching it became a much more boring endeavor to observe the fish!" Keith wrote to me in an email. (She was technically on maternity leave from her post at Lancaster University, but it hadn't seemed to slow the pace of her research, publications, blog posts, and wide correspondence.) "Rather than seeing a couple of aggressive interactions during each five-minute observation," she went on, "we were now regularly seeing nothing happen."

Fortunately, the team didn't let boredom get in the way of amassing a staggering 2,348 such observations on thirty-eight different species of butterfly fish. And all those data pointed in the same direction: when corals bleached, the fish mellowed. Hostile interactions dropped by an average of two-thirds after the bleaching event, turning aggressors into relative pacifists in a matter of weeks. For Keith, this result echoed a textbook prediction about competition over scarce resources. Theoretically—and now apparently in practice—rivals should compete less when their food becomes truly hard to find, because the cost of the fight is greater than the benefit of winning. Bleached corals make a poor meal, and if they die there's no point in eating them at all. So in this calorie-starved environment, butterfly fish become docile in a bid to save energy, a radical behavioral shift that might just let them eke out a living indefinitely, hoping for a return to cooler temperatures more friendly to their favorite corals.

FIGURE 8.3. Aggressively territorial, butterfly fish like this one in the genus *Chaetodon* radically alter their behavior after climate-related coral bleaching events. With food resources scarce, they become meek, directing energy away from fighting toward foraging, dispersal, and other survival priorities. Photo © Elias Levy.

Given the obvious advantages of plasticity, it's worth asking why any species would evolve to lack it. The answer lies in a situation at odds with the current moment: climate stability. During periods of relative calm, which can persist for thousands of years or longer in some habitats, evolutionary pressure often favors specialization. Over time, competition drives some species to find efficiencies—dominating or exploiting particular resources or lifestyles to gain a small but crucial advantage over their neighbors. This often happens at the expense of flexibility—if you want to become a virtuoso on one particular instrument, you can't keep playing every part in the orchestra. The result is an evolutionary tension between specialization (useful when things are stable) and plasticity (useful when things change). The coral-eating butterfly fish embody that push and pull. Mastering the difficult task of digesting stony corals opened up an opportunity and helped them proliferate when times were good, but now that the ocean is warming, that narrow diet leaves them vulnerable. Changing their demeanor so quickly from aggressive to docile was a remarkable feat, but Keith and other scientists see it as a stopgap measure. If bleached corals don't recover, the long-term success of butterfly fish may depend upon an act of even greater plasticity—finding something else to eat.

Similar dilemmas are playing out everywhere, as species struggle to reconcile adaptations that worked well in the past with a new set of rapidly changing conditions. The situation raises an intriguing question, particularly when plasticity falls short. If plants and animals lack a built-in response to climate change, is it possible for them to come up with something new? Can novel traits *evolve* rapidly enough to be relevant to the challenges now unfolding? In 2014, over two dozen top experts reviewed the evidence from hundreds of studies and published their results in a special issue of the journal *Evolutionary Applications*. They concluded that the vast majority of documented climate change responses to date boil down to plasticity—the expression of some latent trait or behavior that a species was already capable of. But in a small and growing number of cases, biologists have begun to identify and measure tangible signs of evolution in action, and one of the most compelling examples lies at the intersection of hurricanes, lizards, and a common backyard leaf blower.

CHAPTER NINE

Evolve

If we want things to stay as they are, things will have to change.

—Giuseppe Tomasi di Lampedusa
The Leopard (1957)

I t's safe to say that Colin Donihue was on a roll. After stints at
Yale and Harvard he landed a coveted postdoctoral position
at France's National Museum of Natural History in Paris. The
research focused on his favorite subject, lizards. And to top it all
off, his field sites were located in the Turks and Caicos Islands, a
popular tourist destination in the Caribbean. During the fall of
2017, he visited two small cays in the center of the archipelago,
where an effort was under way to eradicate invasive, non-native
rats. Among their many crimes, the rats had been preying upon
a unique local lizard in the anole family, a group of small New
World species related to iguanas and chameleons. Donihue and
his team captured, measured, and released scores of the little reptiles, planning to return the following year—after the rats were
gone—to see how the population was faring. (Similar rat removal
efforts had been a boon to lizards on other Caribbean islands.)

Four days after finishing his fieldwork, however, that carefully laid plan went out the window when a raging hurricane slammed directly into his study sites.

"Actually, it was two hurricanes," Donihue clarified, when I called to ask him how the story played out. Hurricane Irma struck first, pummeling the eastern Caribbean with torrential rains, storm surge, and category five winds exceeding 175 miles per hour (280 kilometers per hour). Just two weeks later, Hurricane Maria swept through with similar force. The combined storms devastated low-lying islands like the ones where Donihue's lizards lived, uprooting trees, flattening structures, and leaving both natural and human communities reeling. Needless to say, researchers put the rat project on an indefinite hold. But for Donihue, that setback also offered an opportunity. While his questions about lizards and rat removal would have to wait, he was now in the perfect position to study the effects of hurricanes. Had any lizards survived? If so, and if that surviving population was different from the one he'd just measured, then he might be able to document natural selection in action.

"I thought it was really a long shot," Donihue admitted. But from the standpoint of theory, back-to-back hurricanes made for a powerful evolutionary test. The question was this: Did particular traits help lizards make it through a windstorm alive? If the answer was no, and survival was simply a matter of chance, then surveying the population again would be a waste of time. But if the answer was yes, then Donihue's team had a shot at identifying those traits and watching them spread through the post-hurricane population. "We had no idea what to expect," he told me, "but I knew we weren't going to get another chance at that kind of data." So he cobbled together some funding, headed back to the Caribbean, and found himself in a sort of scientific déjà vu, repeating the exact same field project he'd just completed six weeks earlier.

"We were on a short timeline, so it was pretty much catching and measuring lizards all day long," Donihue recalled. But he described the trip with obvious pleasure, as if this were precisely how anyone would want to spend their time on a tropical island. In conversation, Donihue's enthusiasm for science borders on exuberance. He comes across as someone who probably keeps working—and thinking—long after other people might quit for the day and retire to the poolside bar. That may be why he recognized the potential value of returning so soon to resurvey his lizards. And it's almost certainly why it occurred to him to bring along a leaf blower.

"The customs officer was very confused," he said, and laughed out loud at the memory of trying to explain the science behind traveling with a large piece of landscaping equipment. "We needed to know how the lizards behaved in hurricane-force wind," he told me. "It was entirely possible that they might run for it, or hunker down in the tree roots." Since watching lizards in a real hurricane was out of the question, Donihue used the leaf blower to simulate one inside his hotel room. By placing captured lizards on sticks and cranking up the blower, he could observe their reactions under a variety of conditions. In a moderate breeze, the lizards scurried around to the lee side of the stick and held on tight. As the wind speed increased, their back legs began to slip, until, as the blower reached hurricane strength, they clung on with only their front feet, letting their bodies sail out behind them like flags, parallel to the blow. For thousands of viewers on YouTube, videos of this experiment provided an amusing glimpse into the world of scientific discovery. For Donihue, watching those windblown reptiles provided a precise explanation of the remarkable patterns he was finding in his post-hurricane data.

They started crunching the numbers on the last night of the trip, and it was immediately apparent that something was up. The

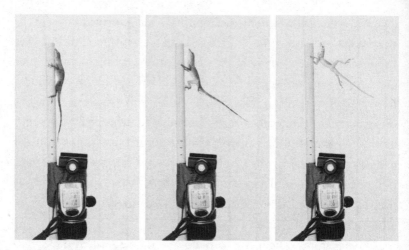

FIGURE 9.1. In a moderate 35-mile-per-hour (56 kilometers per hour) breeze from a leaf blower (left), a Turks and Caicos anole lizard (*Anolis scriptus*) hunkers on the lee side of its stick and holds on tight. Increase the wind speed to 64 miles per hour (103 kilometers per hour) (center), and the back feet start to slip. At hurricane force (right) (84 miles per hour/135 kilometers per hour) the back legs trail out behind like a flag. This posture helps explain why hurricane survivors have stronger front limbs with larger toe pads for better gripping, combined with shorter back legs for reduced drag. (A soft net safely caught any lizards blown free of their sticks during this experiment, and all participants were returned to the wild unharmed.) Photos © Colin Donihue.

surviving lizards, those who managed to cling tightly to trees and shrubs throughout the two storms, had significantly larger toe pads and longer front legs—traits that favored just the kind of gripping power revealed by the leaf blower experiment. What's more, their rear legs were shorter, apparently to help reduce drag in the highest winds, when their bodies were flapping out behind them. Later, Donihue and his team ran various statistical tests that confirmed the evidence was solid. In just six weeks, the lizards in their study population had clearly changed through natural selection—shifting in favor of individuals with beneficial traits. In other words, survival of the fittest.

Donihue called it remarkable to learn that hurricanes were capable of driving evolution, but what really surprised him were the things he discovered next—because his curiosity was hardly satisfied by a single notable finding. Like all good science, Donihue's research is an ongoing process—one question leading to more questions, each new revelation building upon the last. The first thing he wanted to know was whether or not the changes would be passed down. If traits for gripping power weren't heritable, then it wasn't much of a story. So the following year he packed his bags yet again and returned, and once more six months after that, repeating the catch, measure, and release routine that must have had him on a first-name basis with just about every lizard on his study islands. Both trips produced results that were unambiguous: young lizards had clearly inherited large toe pads and other hurricane-ready features from their parents. But that only raised another question—was it just a brief quirk, or could frequent hurricanes induce a long-term evolutionary trend?

"I've been working on that a lot," Donihue told me, but it wasn't a simple question to answer. Natural selection often makes traits "wobble," varying slightly on either side of some kind of functional average. Big toe pads, for example, might be handy in high winds but pointless or even awkward under more typical conditions. If that were the case, and hurricanes were infrequent, then selection pressure would quickly bring the big toe pads back down to "normal" size within a few generations. For Donihue, the issue became whether hurricanes could induce sustained change, pushing traits consistently enough in one direction, over many generations, to result in a lasting outcome. To figure that out he needed three things: more lizards, more hurricanes, and more time.

To arrive at a solution, Donihue began thinking at a different scale, and that in turn led him to a different scientific discipline.

He reached out to a meteorologist, and together they mapped the history of hurricanes across the Caribbean—quantifying where and how often they had occurred. By comparing that map to the various species and populations of anole lizards spread across the same area, he found a telling pattern. Wherever hurricanes were more common, the lizards had bigger toe pads. Selection for gripping ability did indeed appear to be directional, and it had been happening for a long time, shaping lizard feet wherever the weather exposed them to extreme winds on a regular basis. That meant that Donihue's results from the Turks and Caicos Islands were part of something much larger, and it also put his work right at the forefront of climate change biology. "That's the crux of it," he agreed. By confirming real-time evolution in response to weather, Donihue became one of the first people to show that climate change is not only altering what species do, it's altering what they are.

Colin Donihue told me he has plans for a long-term research program dedicated to hurricanes and evolution. He's eager to explore some of the other responses he observed after Irma and Maria, like the uncannily rapid regrowth of damaged trees and shrubs. Is natural selection favoring wind-adapted plants too? What about insects, birds, or mammals? Already, researchers inspired by Donihue's work have found evidence of hurricane selection in spiders, with inherited traits for aggression spreading rapidly through post-storm populations. (Apparently, mean spiders fare better than friendly ones when the going gets tough.) For young scientists, pursuing such questions offers a career with good job security, because while no meteorologist would blame any particular hurricane on global warming, climate scientists agree that the frequency of the most intense storms is on the rise. It's the same for all kinds of extreme weather—when you put more energy (i.e., heat) into a system, the consequences become more severe. Just turn up the flame under a pot of rice, and a messy version of this lesson will take place right on your stovetop.

Weather extremes offer unusually immediate insights into the process of evolution. They are discrete and high-consequence, and if researchers get the timing right, it's possible for them to measure the impacts on a population within weeks or even days. But climate change triggers more prolonged responses as well, and enough time has now passed for at least some of those trends to become apparent. In Finland, for example, tawny owls range in color from pale gray to a rich, reddish brown. Natural selection favored the gray form in the past, when camouflage or some other color-linked trait offered them an edge during long winters with deep snow. But milder temperatures and declining snowpack have steadily eroded that advantage, causing the frequency of brown owls to rise by nearly 200 percent over the past fifty years. In Scotland, speckled wood butterflies have also measurably evolved, with those on the advancing edge of their range developing larger wing muscles to power longer flights northward into warming, newly hospitable territory. Such case studies are compelling, but biologists consider them only half-finished. To meet the gold standard for documenting evolution, observed changes in the wild should be matched to the underlying genetic shifts that drive them. That's no mean feat, but it's increasingly possible thanks to advances in tools used for DNA analysis. Peter and Rosemary Grant's team at Princeton University recently pulled it off in their long-term study of Galápagos finches, linking a gene responsible for beak shape to patterns of adaptation, selection, and even speciation among the birds made famous by Darwin. (Though the Grants' work did not focus on climate change specifically, it's telling that the most dramatic natural selection events in forty years of observation were both weather related— an unusual rainy spell, and a two-year period of drought.)

FIGURE 9.2. Mild winters and declining snow cover in Finland have led to a measurable shift in plumage color for tawny owls, with the once-rare brown form becoming increasingly common at the expense of the gray. *Natural History of Central European Birds* (1899).

As evidence for climate-driven natural selection mounts, there are parallel research efforts delving into other, less heralded avenues of evolution. That's because "survival of the fittest" is far from the only way that species evolve. Mate choice also plays a role, a process Darwin referred to as sexual selection. The concept hinges on attractiveness, the idea that individuals select their mating partners based on particular, identifiable characteristics.

Once that bias is established, competition among potential suitors can spur the rapid evolution—even exaggeration—of those desirable traits. Breeding plumage in birds may be the best-known example, with elaborate male regalia developing in everything from peacocks to roosters to the drakes of various ducks. For songbirds, recent migratory changes suggest that climate patterns are making sexual selection even more important. All across Europe, males of the flashiest species have become the first to take advantage of warmer springtimes, arriving earlier at breeding sites to jockey for position—in effect, extending the season for competition and display. But if the collared flycatcher is any indication, feathers aren't necessarily getting fancier. Sexual selection is very much a two-way street, and on an island in the Baltic Sea, climate change isn't making flycatchers more flamboyant. It's making them dull.

Viewed from the front, the white forehead patch on a male collared flycatcher looks a bit like a paper crown, with short fluffy plumes that stand out in brilliant contrast to the black eyes, bill, and head. Females pay close attention to this feature, and researchers have paid close attention to the results. Beginning in 1980, detailed observations of a population on Sweden's Gotland island showed that males with larger, showier forehead patches earned more mating opportunities and produced more offspring. In short, the females preferred them. Recently, however, that trend has completely reversed. For reasons yet to be determined, rising spring temperatures have made forehead patches suddenly unappealing, or perhaps too costly for the males to maintain (large patches also trigger more confrontations with rivals, and fighting uses more energy in hot weather). Either way, males with flashy foreheads are now reproducing less, leading to smaller and smaller patch sizes with every generation. It's a surprising evolutionary about-face, but one that many biologists see as part of a broader trend. Sexual selection may be driven by attraction, but in the end it boils down to simple economics. Spending energy on extravagant ornaments only pays off if the benefits (i.e., more offspring) outweigh the costs, and fierce competition makes the margins razor thin. If stress from climate change upends that balance, once-positive features can quickly turn into a hindrance—impairing survival, reducing reproduction, or, at the very least, becoming a wasted investment. That is the case for a fish called the three-spined stickleback, whose males combine bright red bellies and blue eyes with rapid zigzag swimming to attract the attention of females. In a warming ocean, however, cloudy, algae-rich waters have obscured visibility at many coastal mating

FIGURE 9.3. Male collared flycatchers with large forehead patches once attracted more mating opportunities, but times have changed, and patch size is now on the wane. Photo © Anton Mostovenko.

grounds, rendering such displays increasingly beside the point. Colors and zigzag swimming are predicted to fade quickly—after all, why dress up and dance for someone who can't even see you?

It takes a lot of data (like forty years of flycatcher observations) to show a definitive change in sexual selection, and the results are often tangled up with competing effects from natural selection, particularly when mates continue to favor a trait that has become detrimental in other ways. But there is another evolutionary force that may be even harder to measure: random chance. Sheer happenstance impacts evolution too, particularly in small populations. Perhaps the best (and tastiest) way to demonstrate this principle involves a bowl of M&M's, or any other colored candies. Take a generous, double handful from the bowl and you're likely to get some of every color. That random mix of colors you selected is analogous to how a large breeding population passes along genetic diversity to the next generation. But if you allow yourself only a small handful of candy, that little group is much more likely to look different from the larger population in the bowl. Some colors might be rare or missing, allowing other colors to dominate—not through adaptation or fitness, but by the simple luck of the draw. Biologists call such randomness genetic drift, a lottery of inheritance that exists to some degree in any population. But drift becomes far more powerful when populations shrink and become isolated. This is precisely the scenario faced by species with dwindling habitat, or by those dispersing in small groups to colonize new areas. Surviving populations will long bear the mark of their ordeal in the reduced genetic diversity they pass along. (To continue the candy analogy, the descendants of tiny handfuls are unlikely to quickly reinvent lost colors.)

Scientists know that climate-driven genetic drift is happening— it is mathematically unavoidable. But no one has yet isolated its effects from all the other things impacting, fragmenting, and otherwise reducing plant and animal populations. That will take time. Meanwhile, there is another kind of evolution on the rise that produces immediate results. And in one of the best-known case studies, it's possible to hold the consequences in your hand. But only if you know where to drop your line, and what to use for bait.

The oarlocks creaked as we rowed quietly into a corner of the lake we'd come to know as Trout Bay. It was little more than a slight indentation in the forested shoreline, but it usually produced a re-liable bite on spoons or spinners dragged close to the bottom be-hind a flashy troll rig. Historical accounts spoke of local cutthroat trout the size of salmon, and hauls of more than a hundred fish in a single outing. We weren't having that kind of a morning. Our luck so far echoed an even older narrative, a local Native Ameri-can story about Raven venting his rage at a water spirit by casting thunderbolts into the lake, killing off all of its fish for generations. In the version of the tale I'd heard, nothing was said about what had made Raven so angry in the first place, but I was starting to wonder if he hadn't been having bad luck fishing.

Just then, the tip of Noah's rod gave a telltale tug and for an instant all frustration was forgotten. As he started reeling in, however, that gentle pull never got any stronger, and when the fish came to the surface we saw why. It was a minnow—a juvenile bass not much larger than the lure itself. We laughed, released the tiny fish, and watched it dart back down out of sight. It wasn't the species we'd been hoping for, but then again, we weren't fishing where we'd hoped to fish.

Occasionally in life, work and pleasure coincide as if by prov-idence. When I wrote a book about feathers, for example, the research involved a lot of happy birdwatching; likewise my book on seeds had me investigating favorite seed-based indulgences like coffee and chocolate. So it seemed as if kismet had struck again when I discovered that one of the world's finest examples of climate-driven evolution was taking place in one of its finest trout streams. Noah and I both love to fish, so I immediately began making arrangements for a working vacation to Montana's Flathead River Valley, only to see those plans undone by a different sort of biology. In the spring of 2020, the spread of novel coronavirus meant that fishing trips of any kind would have to remain close to home. Even though our local catch wouldn't include the unusual trout in question, the situation didn't prevent me from at least talking to the scientist in Montana whom I'd intended to visit. His research lay at the heart of the evolutionary puzzle unfolding there, and his eagerness to talk about it helped bridge the distance between us.

"I've basically been obsessed with fishing since I was a kid," Ryan Kovach said at the start of our phone call, and described how the allure of rod and reel had shaped quite a few of his major life choices. Leaving home for college? "I basically chose Montana because I knew the fishing was going to be good." His first major research project? Trout genetics in Yellowstone National Park—more good fishing. Graduate school? Pink salmon in Alaska. And finally, he returned to Montana where he now works as the state geneticist for the Montana Department of Fish, Wildlife, and Parks—and where an online biography quips that "he may undo many positive conservation impacts by tirelessly trying to catch all manners of fish . . . but cannot stop himself from doing so."

Kovach is hardly the only compulsive angler in Montana, a fact that helped set the stage for his groundbreaking climate change research. "They stocked frightening numbers of rainbows in the Flathead," he told me, referring to the common practice of releasing captive-bred rainbow trout into public waterways. Fish and game departments use this tool throughout the American

West in an effort to boost people's chances of pulling in a keeper. (Stocked rainbows were the fish Noah and I had been trolling for in the island lake near our home.) Unfortunately, this annual flood of hatchery-raised fish has replaced native populations in many lakes and rivers, sometimes through direct competition, but also through the often-overlooked evolutionary process that has become Kovach's specialty: hybridization.

Whenever two closely related species interbreed, large amounts of genetic material can be transferred all at once. In plants, the results often create novel evolutionary pedigrees, and hybrids are considered a major source of new species. Animal hybrids, on the other hand, are usually infertile. Horses and donkeys cross to produce mules, for example, but the story stops there because the mules themselves can't breed. That's not the case, however, when rainbow trout meet the native westslope cutthroats of the Flathead River and other western streams. Not only are their offspring fertile with one another, they cross back with the parent species, creating a pathway for genes from one to steadily infiltrate the other. Experts call this process introgression, and its consequences can be significant and long-lasting. Introgression explains why Neanderthal DNA still shows up in modern human genes for everything from skin pigmentation to hair growth, even though we stopped interbreeding with them over forty-five thousand years ago. For Montana's cutthroat trout, that example also contains a more ominous lesson: while some Neanderthal genes may persist in modern humans, the Neanderthals themselves are long gone.

"It's like a conveyor belt," Kovach said, describing how rainbow trout—and their genes—are swamping Montana's native cutthroats. As temperatures rise with climate change, warm-water rainbows are moving steadily upstream into mountain tributaries that once served as cutthroat strongholds. "They're invading the last of the best habitat," he went on, and explained that wherever

FIGURE 9.4. Westslope cutthroat trout (*Oncorhynchus clarkii lewisi*) in Montana's Flathead River contain increasing amounts of DNA from rainbow trout. Climate-driven hybridization between the species has begun swamping the cutthroat population with genes from their more populous rainbow cousins, whose range is expanding as waters warm. Photo © Jonathan Armstrong.

the two species mingle, they interbreed. The resulting "cutbow" hybrids then carry rainbow DNA even farther into the cold-water refuges, where they breed with the few remaining purebred cutthroats. "Introgression goes beyond thermal limits," he explained, meaning that rainbow genes threaten to overwhelm cutthroats through hybridization, even in places the pure rainbows themselves never reach.

When I asked Kovach why rainbow genes were swamping cutthroats and not the other way around, he told me it was a numbers game. "Rainbows were stocked in the most productive parts of the rivers," he said, to the tune of twenty million fish in the Flathead alone. In addition, they have a stronger tendency to wander as they mature, creating a steady flow of breeding-age interlopers to foist their genes on their less numerous, stay-at-home cousins. "The irony is, if it weren't for rainbows, cutthroat trout might actually be benefiting from climate change," Kovach said, citing unpublished data from one of his graduate students that showed cutthroat productivity otherwise increasing in warmer mountain streams. As it is, Kovach doubts that the Flathead or other western rivers will continue to harbor any pure cutthroat populations, just "residual bits" of cutthroat DNA, tucked away in fish that look and behave like rainbows.

The results of Ryan Kovach's research must strike him as bittersweet—it tells a fascinating story biologically, but comes with the knowledge that a species will ultimately disappear, one that he will miss catching as well as studying. Still, Kovach was quick to point out that hybridization isn't always a negative force in evolution. Plant hybrids are often fitter than their parent species, at least initially, and even in fish there are examples of inbred populations benefiting from the influx of new genes. For species in decline, hybridization can sometimes preserve unique genetic material that would otherwise be erased by extinction (like those Neanderthal traits hanging on in *Homo sapiens*). The implications vary case by case, but with climate change causing so many species to shift their ranges and bump into each other in new ways, one thing is abundantly clear: the hybrid caseload is rising.

Before giving up on our fishing expedition and consoling ourselves with ice cream, Noah and I rowed out from the shoreline and dropped our lines down to the bottom. If the trout weren't biting, there was always the chance for kokanee, a type of land-locked salmon found only in the deepest, coldest parts of the lake. That habitat seemed destined to shrink as summertime tempera-tures warmed, but if our fishing hole was deep enough to remain cool, its kokanee might just find themselves in an enviable posi-tion. By various quirks of landscape (or waterscape), some species will be able to cope with climate change by continuing life pretty much as normal. But for that strategy to be effective, they must rely on an axiom borrowed from the real estate industry: location, location, location.

CHAPTER TEN

Take Refuge

To change and change for the better are two different things.

—German proverb

For graduate students anywhere in New England, an autumn day in the woods can't begin properly without cider doughnuts. The local recipe involves mixing a generous splash of fresh apple cider right into the dough and topping the finished product with ample amounts of cinnamon and sugar. Paired with strong coffee, these seasonal treats revitalize brain waves dulled by long hours and late nights of coursework and research. My colleagues and I had secured a fresh supply in the town of Bristol, twenty-five miles south of our base at the University of Vermont, and conversations in the van were already perking up as we drove on into the nearby hills. It was a Friday, a time when all of us getting our master's degrees in the small Field Naturalist Program put other tasks on hold, and headed outside for some hands-on experience. We toured bogs and marshes, inspected quarries and rocky outcrops, and hiked everywhere from lakeshores to mountaintops, all in the company of a rotating cast of experts. The trips emphasized hidden connections in nature, how everything from bedrock and soils to weather patterns and local history can influence the plants and animals found at a given site. We learned something new every week, but none of those lessons changed my thinking about landscapes—and climate—more than the mystery we encountered at the base of Bristol Cliffs.

Our guide for the day was Alicia Daniel, herself a graduate of the same program, who would go on to hold the enviable job title of official naturalist for the city of Burlington, the largest town in the state. She already had an intimate knowledge of local terrain and ecology, but was keeping conspicuously quiet about our destination as the van rattled up a narrow back road. We pulled over and parked on a steep slope below the rugged face of the cliffs, a series of sheer drops and ledges near the summit of the ridge above. Daniel led everyone uphill through a fairly typical hardwood forest dominated by sugar maples, Vermont's state tree. But soon the ground became decidedly lumpy, and our true goal appeared through the thinning vegetation—a barren jumble of blocks and boulders at the foot of the cliffs known to locals as Hell's Half Acre. In fact, the talus slope covered nearly forty acres, an expansive accumulation of slides and rockfalls from the upper reaches of the precipice. The pieces ranged in size from ankle-turning cobbles to slabs larger than houses, and my notes from the trip tell me they consisted of something called Cheshire quartzite. But we hadn't come to Bristol Cliffs to focus on its geology, at least not directly. The real puzzle lay in what that brittle stone had enabled, through the architecture of its haphazard erosion.

It was a warm day for October and we'd all worked up a bit of a sweat climbing the hillside. So my first impression at the base of the talus was one of refreshing, shady coolness. Then I noticed the trees. Instead of leafy hardwoods, we stood suddenly surrounded by conifers—red spruce, black spruce, and balsam fir. Looking down, I found the groundcover equally transformed, with a sudden profusion of cold-hardy shrubs like Labrador tea and leatherleaf. There was even reindeer lichen and a patch of *Sphagnum* moss. Somehow, in a matter of steps, we seemed to have traveled hundreds of miles northward to the boreal forests of Canada, or ascended two thousand feet (six hundred meters) in elevation to the frosty heights of the nearby Green Mountains. Backing just a few yards away from the talus produced the reverse effect, an instant return to hardwoods. After giving everyone time to explore this odd dynamic, Daniel gathered the group back in the conifer grove to talk about what we'd seen.

Sitting on the ground, or perched on mossy rocks, we cooled off quickly in the oddly chilly air, and people started donning jackets and searching through backpacks for sweaters and hats. Temperature lay at the heart of the day's story, and I learned later that Daniel always sat her students in the same place when she visited Bristol Cliffs, letting the physical reality of the place subtly reinforce the academic lesson. It's the sort of attention to detail that makes her a good teacher, but we still needed a few more clues to figure out what was going on. Clearly, the conifers were growing in a cold pocket, but the situation seemed odd so close to the obvious warmth of the talus field. Like a sunbaked parking lot, it radiated heat as the day wore on, and we could see hawks and turkey vultures circling on the thermals rising above it. The whole slope faced west, catching the sun's rays at peak strength and making conditions, if anything, warmer than usual for Vermont. Oaks, hickories, and other species more typical of southern forests thrived amongst the maples. Yet something made that one tiny conifer patch, nestled right against the toe of the slope, look and feel like a habitat from the far north. Finally, Daniel directed our attention not to the rocks of the talus, but to the spaces in between—how the cobbles and boulders leaned against

FIGURE 10.1. Talus slopes in New England capture cold air amongst their cobbles, creating conditions for conifers and other boreal forest species to persist within a landscape dominated by hardwoods. Illustration © Libby Davidson.

one another to form a vast network of tiny caves and crevices. Up close, we could feel a cold breeze wafting outward from those dark gaps, and when someone reached into a particularly deep alcove, they found a chunk of ice.

"It's basically refrigerated air," Daniel told me recently. I called her at home in Vermont—two decades after our first meeting—to hear her explain again how cold, dense air can sink through a talus slope and trickle out at the base, creating its own tiny climate. While the surface of the rocks may heat up on a sunny day, that energy never penetrates to the shady depths below, and nighttime always brings a fresh chill as the rocks and surrounding landscape cool. At Bristol Cliffs, the effect is enhanced by ice that fills the deepest cracks during winter and remains frozen for much of the year, and by a shelf of bedrock beneath the talus that funnels the downdraft to a site suitable for trees and shrubs. The result is a habitat in miniature, a patch of cold ground Daniel estimates as "about the size of a swimming pool." Beyond that zone the chill dissipates, but within it grows a flora that is distinctly out of place. And, just as remarkably, out of time.

Turn back the clock far enough in Vermont, or anywhere else in New England, and you wouldn't need a talus slope to find refrigerated air. Eighteen thousand years ago, the entire region lay under a continental ice sheet that stretched from the Arctic south to the vicinity of modern-day New York City. After the ice retreated, tundra plants established first, followed by a boreal forest that blanketed the landscape for over 2,500 years. As the climate continued to warm, however, those conifers shifted steadily northward to be replaced by hardwoods. That is to say, most of them shifted. Wherever things remained cold enough—be it a mountainside or a peculiar patch of talus—they did something else. They took refuge. It is entirely possible, even likely, that the handful of spruce, fir, and other boreal species at the base of

Bristol Cliffs have persisted there for thousands of years, generation after generation, making the most of that trickle of chilly air
while the forest all around them warmed up and changed. The
alternative explanation requires an implausible series of long-
distance dispersal events, with the seed or spores of every northern species in the grove traveling scores of hundreds of miles and
just happening to land on that one suitable fraction of an acre.
Either way, the lesson is one of cause and effect: plants and animals respond to the conditions their landscape provides, no matter how atypical or idiosyncratic. In the context of adjusting to
climate change, places like Bristol Cliffs provide their few lucky
residents with an appealing option: business as usual.

"Whatever was growing there was always atypical," Daniel
mused, when our conversation turned to the postglacial history of
the talus slope. Tundra species probably persisted at Bristol Cliffs
long after conifers moved into the neighborhood, just as the conifers are now embedded in hardwoods. It's not as if the place
is impervious to warming, but the cold air acts as a buffer, slowing what biologists refer to as the velocity of a changing climate.
The Bristol Cliffs talus is an extreme case, but similar principles
apply wherever conditions remain abnormally cool—deep, shady
valleys, for example, or north-facing slopes with limited direct
sunlight (south-facing below the equator). In freshwater systems,
cold springs and snowmelt can have a similar effect, just as cool
currents or deepwater upwellings can in the ocean. When such
anomalies last, they allow the species within them to endure as
small, isolated populations in the midst of an otherwise inhospitable environment. The future may catch up eventually, but if the
time lag is long enough, and if climate trends stabilize or reverse,
then history suggests these places don't just slow the effects of
climate change, they help plants and animals survive it.

In science, few things signal enthusiasm for an idea more than the invention of a new word to describe it. Accordingly, *refugium* was coined specifically for places like the Bristol Cliffs talus, where species can find shelter when conditions elsewhere become unfavorable. The term, which first appeared in 1902, was used to reference several deep mountain lakes in Switzerland, where cold water had allowed various northern fishes and crustaceans to persist since the last ice age. From the beginning, then, biologists have linked this concept to surviving climate trends—not just warming, but also cooling, drying, or any other significant shift. Soon after our trip to Bristol Cliffs, my Vermont colleagues and I saw the reverse pattern far to the north in Quebec: a patch of maples and oaks surrounded by boreal conifers. They grew where a south-facing cliff trapped and reflected heat to nearby trees, and were considered relicts of a warm period several thousand years earlier, when hardwoods had briefly edged north of their current range. Well-studied postglacial cases of all kinds exist throughout Europe and North America, and people have also argued for the importance of refugia in the tropics. In Africa's Congo Basin, for example, intermittent dry periods occurred throughout the Pleistocene epoch (2.6 million to 10,000 years ago), repeatedly shrinking the region's great rainforest into a patchwork of remnants—refugia—separated by savanna grasslands. Forest species concentrated in those residual stands later spread out again when conditions improved, but often at different rates, and sometimes in forms slightly altered by their isolation, creating patterns still visible in the modern distribution of everything from snails to primates. Gorillas, to take one famous example, never fully recolonized the central portion of the basin after its reforestation, and now occur as two distinct species groups, one on its eastern edge and one over six hundred miles (one thousand kilometers) to the west.

Researchers interested in refugia devoted most of the twentieth century to looking backward, studying how species had survived past ice ages and other environmental upheavals. But the pace of recent climate change has given the idea sudden urgency, shifting the focus firmly to the future. Where will quirks of landscape, water temperature, or weather slow the velocity of modern climate change, and what species will be on hand to find refuge? Alicia Daniel didn't have a ready answer when I asked her what might be growing below Bristol Cliffs in a hundred years. She paused to think, and I imagined her flipping through a mental Rolodex of local tree species, the cards well thumbed after decades of constant use.

"The trees people are most concerned about around here are sugar maples," she said finally, noting how sap flow in springtime was already being disrupted by increasingly chaotic weather. That not only affected the health of the trees, it threatened production of a seasonal sweet even more important to New Englanders than cider doughnuts—maple syrup. Industry researchers predict that "the region of maximum sap flow" will shift hundreds of miles northward as the climate warms, and savvy Vermont producers are already hedging their bets by tapping trees in the coldest parts of local forests. Management guidelines presented at the Vermont Maple Conference in 2017 referred to such places with a familiar word, refugia. So it's not at all far-fetched to imagine a few sugar maples one day finding shelter at the Bristol Cliffs talus, replacing the conifers and hanging on in a forest of oaks as the whole landscape shifts to a warmer regime. Perhaps they will even have buckets dangling from their trunks, tapped by some enterprising neighbor longing for a nostalgic taste of local syrup.

Like musing about maples at Bristol Cliffs, most studies of modern climate change refugia involve a certain amount of educated guesswork. They're based on expectations rather than outcomes,

and so include a lot of terms like "assessment," "future vulnerability," and "conceptual framework." For biologists, the idea is to
pinpoint and ultimately protect places that could serve as safe ha-
vens for biodiversity, and various predictive models have already
identified a wide range of promising sites. Cold mountain streams
in the American West, for example, are expected to warm slowly
and harbor native trout and frogs for many decades, while in Australia's Eastern Highlands, shady sites with shallow groundwater
promise to shelter a range of plants and animals from increasing
droughts and fire risk. One ambitious project in Sweden identified
ninety-nine precise locations where boreal forest species persisted
south of their core range, and then measured climate variables at
those sites eight times a day for a full year. They found that even
subtle differences in temperature and sunlight provided a buffer
from climate change, regardless of how small the area (thus another new word, *microrefugia*). Huge uncertainties remain, and
most studies conclude with as many questions as answers. How
large does a refugium need to be? How long can isolated populations remain viable? Is it more important to maintain average
temperatures or to reduce extremes? How critical is moisture?
What about vital interactions like pollination or predation? Some
experts doubt that refugia will harbor enough species, for a long
enough time, to make any difference at all. ("They aren't going
to cut it," marine biologist Gretta Pecl told me dismissively.) But
with conditions changing so quickly, it's now becoming possible to do more than make predictions. For at least one species in
the mountains of the American West, taking refuge has already
proven to be an effective and lifesaving response.

If you can imagine a grayish brown, rabbitlike creature the size
of a grapefruit and nearly as round, that is an American pika.
They inhabit mountain slopes, often well above tree line, from

the high Rockies west to the Pacific Ocean. Slow to breed and reluctant to disperse, pikas have long been considered at great risk from rising temperatures—like other alpine residents, they have no place to retreat to when habitats begin shifting uphill. But new research has opened a window of hope, because pikas benefit from the same principles at work in places like Bristol Cliffs. They live almost exclusively in and around talus slopes, nesting in crevices between the rocks, and venturing out only a few feet to gather grasses and wildflowers from nearby meadows. (They drag the snipped vegetation back home for later consumption, storing it in tidy piles charmingly referred to—even in scientific papers—as haystacks.) Because of the way cold air gathers in talus, pika habitats in places like California's Sierra Nevada mountains average 7 degrees Fahrenheit (3.8 degrees Celsius) cooler than their immediate surroundings during the summer. And just like at Bristol Cliffs, that cold air seeps out into adjacent vegetation at the base of the slope, often helping to maintain a patch of the very alpine meadow plants that pikas prefer. With these effects in mind, a team led by US Forest Service ecologist Constance Millar began resurveying large swathes of potential habitat—not just in alpine areas, but also where talus occurs in places far warmer than anyone had generally thought to look.

"The first thing I realized was that I needed contact lenses," Millar told me with an easy laugh. "I couldn't see the pellets!" Because pikas spend so much time hidden amongst the rocks, the best way to locate them involves spotting their tiny, distinctively round droppings. Once she had the proper eyewear, however, that task proved surprisingly easy. "We started finding a lot of populations," she said, and explained that many of them turned up at lower elevations, in talus surrounded by habitats like pine forests or even sagebrush. For the pikas, all that seemed to matter was a nice, cool rock pile with a bit of meadow at its base. For Millar

FIGURE 10.2. Mountain-dwelling American pikas (*Ochotona princeps*) are buffered—at least temporarily—against the effects of a warming climate by their habit of living in and around talus slopes, where cold air gathers and makes their habitat slower to change. Photo © Bryant Olsen.

and her team, what mattered was that climate change refugia had suddenly been transformed from theory to reality. Pikas were already taking advantage of them, and—by all indications—they had been for a very long time.

"They're cold-adapted. Historically, pikas flourished during glacial periods, not warmer times like now. They even used to live in the lowlands!" Ideas tumbled out of Millar in a rush, and my note-taking hand was cramping as I tried to keep up. She seemed eager to pack as much information into our phone conversation as time would allow, a habit that probably helps explain her prolific career. Over the course of four decades, studying everything from pikas to bristlecone pines, Millar had achieved what a profile in the *New Yorker* magazine recently called revered status among mountain biologists. She talks quickly, but she's also a good listener, and when I told her about my long-ago visit to Bristol Cliffs, she wanted to know exactly where it was so she could "put it on her list" of places to visit. The spruces and firs eking out a living there struck a chord with her because pikas have been doing the same thing—using talus slopes as refugia throughout the long, steady warming since the last ice age, just as they must have done repeatedly during the topsy-turvy glacial history of the Pleistocene. Modern climate change may have shifted that process into high gear, but the pattern is an ancient one.

Living within refugia is a lucky break, but it doesn't mean that pikas have nothing to worry about. They remain highly sensitive to heat stress, loss of snowpack, and other climate-related changes, and scores of populations have indeed winked out in recent decades. "Their range is going to shrink. There's no question about that," Millar agreed, and even the best talus sites may eventually "warm out" if temperatures rise unabated. She paused for a second, and then added a comment that managed to capture both the limits of refugia and their potential. "They're just buying time," she said, "but probably a lot of time."

On a small scale, relationships between landscape and climate can be experienced anywhere. Our garden, for example, receives plenty of sunlight all day long, but it sits in a shallow depression where cold air settles after dark, and so we've never been able to reliably grow and ripen tomatoes outside the greenhouse. Just down the road, a small hill provides our neighbors with a south-facing slope that is warm enough for tomatoes and other heat-loving crops to thrive. Similar principles apply a few miles away in town, where I recently gathered some data on a bright February afternoon. The main street runs from east to west, which, given the low angle of winter sunlight, puts the sidewalk on the south side in the shade of buildings from roughly mid-December through March. Not surprisingly, I found temperatures there 5.5 degrees

Fahrenheit (3 degrees Celsius) cooler than those directly across the street to the north, where the sun was shining and reflecting down off walls and windows. The landscaping was responding accordingly. On the sunny north side, twinberry shrubs had leafed out, the Oregon grape was blooming, and early spring wildflowers like crocus and iris were already in their full glory. Across the street, everything was still dormant and winter bare. (It's also worth noting that 65 percent of the pedestrians I encountered had chosen to walk on the sunny side of the street. Plants aren't the only organisms that appreciate a warm winter microclimate.) Contrasts like these are daily reminders that climate plays out unevenly in all but the most homogeneous landscapes. Such inconsistencies exist all around us, but they only qualify as refugia when the differences are acute enough to generate a distinct environment, and persistent enough to put that environment on a different climate change trajectory. Talus and other quirks of topography can make that happen, but perhaps an even more straightforward way occurs in the ocean, where one set of environmental conditions can be literally transported, fully formed, into the midst of another one. That's exactly what happens along certain coastlines, where dense, cold water from the depths surges to the surface in plumes known as upwellings. Like talus slopes, they're another place where at least some species have already begun taking refuge.

Portuguese marine biologist Carla Laurenço learned about upwellings accidentally, during her doctoral research, when she noticed an intriguing pattern while surveying tide pools and rocky shorelines from the Western Sahara north past the Strait of Gibraltar to the Iberian Peninsula. She described the upwelling process to me as an interaction between wind patterns and coastal geography. When strong, steady winds consistently push surface water away from a single location, deep water rises up to fill the void. Local fishers know all about such places because they're particularly productive, their food chains boosted into overdrive by nutrients brought up from the depths. For Laurenço, those cool, rich waters drew her attention to a species she hadn't even set out to study.

"The results were quite unexpected," she wrote in an email, and explained that much of her work focused on the genetics and distribution of invertebrates like mussels. But it quickly became apparent that something was going on with a type of brown algae that typically shared the same intertidal rocks. Known as bladder wrack or rockweed, it belonged to the genus *Fucus*, a dominant group along rocky shorelines throughout the northern latitudes. All *Fucus* species have branched, flattened blades that lie limp on the rocks at low tide, but sway gracefully in high water, held several inches aloft by tiny, built-in floats. (Pinching these little pockets produces a satisfying snapping sound—when I was young we called the local variety "pop-weed.") The *Fucus* in Laurenço's study prefers cool water, and climate-driven warming along the coast of northwest Africa—a marine hotspot—has driven it steadily northward. But in five distinct upwelling zones, where peak water temperatures remain as much as nine degrees Fahrenheit (five degrees Celsius) colder than their surroundings, Laurenço found *Fucus* populations not just surviving, but thriving. Even their genetic diversity was high, and it showed patterns of past isolation, as if this were far from the first time that *Fucus* had used such sites as refugia.

After our email exchange, I took a copy of Laurenço's paper with me and read it again at the southeastern tip of the island where I live, a place where our own local *Fucus* shrouds the rocky point like an umber blanket. As yet, our *Fucus* shows no signs of beating a climate-driven retreat, so the comparison to Laurenço's experience was imperfect. But I wanted a better sense of what it

FIGURE 10.3. This rockweed (*Fucus guiryi*) continues to thrive in cold-water refugia along the northwest coast of Africa, adding structure and shelter and sustenance to a diverse intertidal community. Photo © Carla Laurenço.

must have been like for her to stumble across those upwelling refugia on the African coast and suddenly encounter lush patches of *Fucus* living an untroubled existence.

The tide was falling, and visible currents swirled through the deep channel beyond the point, kicking up lines of standing waves that darkened the water like windrows. I clambered across the slick rocks to the nearest patch of *Fucus* and couldn't help squeezing one of its pale, brown bladders between my fingers. It popped with an audible crack, just like old times. Laurenço described *Fucus* as a foundational species, something that harbored a diversity of other life-forms within its tangled canopy. Sure enough, when I parted the brown fronds I found the rocks beneath festooned with limpets, periwinkle snails, blue mussels, and the vivid, mauve crust of a coralline algae. Nearby, a flock of black turnstones foraged, ankle-deep in the brown tangle, probing for crustaceans. A pair of song sparrows soon joined them, lured down from the forest by the promise of an intertidal meal. Everything looked perfectly normal, the same kind of rocky shoreline scene I remembered from childhood. And that is the appeal of refugia—the idea that some places will remain unaltered in the face of rapid change. But here Laurenço's study is particularly revealing, because while she encountered a few other cold-water species alongside the *Fucus*, there were dozens of things missing. Instead of preserving a complete intertidal community, the upwellings appeared to contain only a smattering of its original residents. Conditions that suited *Fucus* didn't necessarily satisfy all of its neighbors, a reminder that refugia are inherently erratic in the species they happen to preserve. Among responses to climate change, taking refuge may amount to less of a remedy than a happenstance.

Of course, chance plays a role in any climate outcome, and that's where this narrative now turns. Biologists may study how plants and animals respond to rapid change in the here and now, but their work invariably arrives at a question about the future: How does what we've learned about the present (and the past) hint at the nature of what comes next?

PART FOUR

The Results

Fall down seven times, get up eight.

—Japanese proverb

As an author, I've learned that certain questions arise repeatedly over the course of a long project. When I was writing a book devoted entirely to feathers, for example, people would ask me, "How is that possible?" When I turned my attention to bees, everyone wanted to know how many times I'd been stung. And now that I'm focused on climate change, I've come to expect questions about the future—"What's going to happen?" Nobody knows the full answer, of course, but facets of it lie in the many biological challenges and responses we have already explored. For some scientists, however, measuring those observable changes is just a starting point. In this section we investigate the potentials and pitfalls of prognostication, how models are made, how surprises are certain, and how the clearest signs of what to expect may lie in what has already happened . . .

CHAPTER ELEVEN

Pushing the Envelope

*But who wants to be foretold the weather? It is bad
enough when it comes, without our having the misery of
knowing about it beforehand.*

—Jerome K. Jerome
Three Men in a Boat (1889)

I woke to the sound of water—rushing river noises, and the
steady drumming of raindrops on a metal roof. Gradually, my
senses also informed me of pale dawn light at the window, rising birdsong, and, of course, the smell of socks. Descriptions of
tropical research rarely mention such odors, but they're an unavoidable consequence of hot weather, strenuous fieldwork, and
scarce laundry facilities. The fragrance becomes particularly acute
if you're sleeping dormitory-style in a stuffy room full of strangers,
with wet clothes and equipment hanging from every available peg
and bedpost. As a frequent visitor to La Selva Biological Station
in Costa Rica, I might have qualified for a housing upgrade. Some
of the better facilities featured verandas, private baths, and even
air-conditioning—nice enough to attract a few tourists mixed in
with the regular stream of scientists and students. Yet every time

I made a reservation, I asked for a bunk at River Station, the oldest and most isolated building on campus. Call it due regard, or even superstition, but I wanted to stay under the same roof where a great scientific idea had taken shape—one that happens to have paved the way for nearly every climate change prediction in biology.

Dressing quietly, I crept from the room before anyone else was awake. At field stations like La Selva, it's often possible to guess another scientist's discipline by what time they show up for breakfast. Ornithologists, for example, rise early with the birds, but my current roommates were clearly studying something nocturnal—dragging in late every night, and rarely stirring before ten o'clock in the morning. My own project focused on trees, which are obliging subjects at any time of day. But I'd gotten up especially early for something unrelated. I wanted to investigate a hunch that I had about Leslie Holdridge, the tropical forester who'd built River Station as his personal research retreat in the early 1950s. He called the surrounding property *Finca La Selva*, "The Jungle Farm," and used it for experiments on tree crops like cacao and peach palm, interplanting them with native species as an agricultural alternative to clearing the rainforest. But that wasn't Holdridge's only forward-thinking idea. His time at River Station also produced the definitive version of what came to be known as the Holdridge Life Zone System, a way of combining simple climate variables to predict the vegetation and habitat conditions for any location on the planet.

Outside, I slipped my feet into knee-high gumboots and paused to look around. River Station resembles a cross between an army barracks and something by Frank Lloyd Wright, with a wood-paneled second story cantilevered out over the first to form a covered walkway. It sits on a bluff above the Sarapiqui River, an audible but invisible presence through the wall of dense greenery pressing in on all sides. In that regard, the setting seemed oddly claustrophobic for the contemplation of such a global theory. Lowland rainforests like the one at La Selva vary a lot more in their biology than in their scenery, and Holdridge himself described the unrelenting vegetation as overwhelming. Yet even in such thick jungle there was a way to put his ideas about climate and habitat to the test. After a quick stop at the dining hall, I headed for a trail that crossed the southwestern edge of the property, where, if my suspicion was correct, I could hike into a completely different life zone, and be back in time for lunch.

The path took me past laboratories and classrooms and then deep into La Selva's four thousand acres of primary and regenerating forest. In Holdridge's day, similar vegetation had stretched for

FIGURE 11.1. For Leslie Holdridge, the hot, wet rainforest at La Selva Biological Station in Costa Rica embodied the effects of climate on vegetation. Photo © Thor Hanson.

miles all around, but now La Selva was a remnant. It contained one of the largest patches of trees left on the coastal plain, surrounded on three sides by pastures, banana plantations, and broad fields of pineapples. The fourth side of the property, however, did not border the lowlands. It ran smack into the toe of Costa Rica's Cordillera Central, a range of high volcanic mountains that rose up suddenly from the flatlands like a dark wall. Luckily, the neighboring property was also protected, part of a national park that made it possible to hike from near sea level at La Selva all the way up to the summit of Volcan Barva, towering above at over 9,500 feet (2,900 meters). While it's unclear if Holdridge ever walked that particular route, he certainly would have known what to expect. He based his life zones on how plant communities responded to climate, and—as scientists had known for over a century—few places made that relationship more clear than a big, tropical mountain.

Epiphanies rarely stand alone, and Holdridge's breakthrough was no exception. It was built directly upon the work of nineteenth-century German naturalist Alexander von Humboldt, who explored Mount Chimborazo in Ecuador and mapped out the distinct bands of vegetation that grew along its flanks. At first glance, Von Humboldt's diagram looks self-evident: habitats and species change as you climb up a mountain. But read the small print and it's clear that he understood something more fundamental. Elevation was secondary. *Climate* dictated what grew where, and similar vegetation should occur in any place where temperature, humidity, and other conditions were the same, regardless of geography. Trees, for example, don't survive in places where temperatures average less than forty-two degrees Fahrenheit (six degrees Celsius) during the growing season, a situation that occurs at 11,600 feet (3,550 meters) on Chimborazo, at 7,200 feet (2,200 meters) in the Swiss Alps, or at a few feet above sea level

FIGURE 11.2. Alexander von Humboldt's famous illustrations of Mount Chimborazo in Ecuador explored universal relationships among climate, elevation, and habitat. This detail shows bands of vegetation and corresponding species names, ascending the flanks of the mountain. *Essay on the Geography of Plants* (1807). Zentralbibliothek Zürich/Public Domain.

in Northern Canada or Siberia. In all of those disparate locations, forested habitat gives way to tundra as soon as that threshold is crossed.

When Leslie Holdridge turned his attention to the idea of life zones, he found a research topic surprisingly little changed since Von Humboldt's day. Other attempts to classify the relationship between climate and habitat had strayed from what Holdridge saw as Von Humboldt's basic insight. Starting with similar variables, and working in a similar tropical landscape, Holdridge devised a model based on three simple measurements: heat, precipitation, and moisture. For the first, he used what he called biotemperature, a metric that reflected the period of time when plants were actively growing. He drew on standard rain and snow measurements for the precipitation data, and the final variable involved a combination of the first two—capturing a biologically relevant indication of the moisture available to plants. (It's worth explaining that Holdridge's botanical focus didn't just stem from his bias as a forester—terrestrial ecosystems are almost always defined by the plants that give them structure, not the animals that live there. Better to call a forest a forest, for example, than a "bird-and-squirrel area, with trees.")

To illustrate his system, Holdridge arranged thirty of the most prominent life zones into a triangular grid, like a habitat Ouija board, ranging from ice and tundra at the peak to hot, tropical zones at the base, and with wet forests on the right-hand side transitioning through savanna and scrub to arid deserts on the left. Though its tightly packed cells and dotted lines lacked the aesthetic appeal of Von Humboldt's Chimborazo diagram, Holdridge's triangle neatly captured the underlying relationships. Textbooks still use this triangle to explain the influence of climate on habitat. But for Holdridge, the diagram was only a crude representation of his idea. He imagined the system in three dimensions, more like a pyramid, with space for hundreds of life zones and subzones, each one a conceptual building block defined by the natural subtleties of climate. That vision may be Holdridge's most prescient and lasting contribution, because advances in computing power have made such complex abstractions not only possible but commonplace—a basic tool of modern biology.

As the trail climbed steeply up from La Selva into the foothills, it became clear to me why Leslie Holdridge had spent so much time thinking about heat and moisture. I felt like I was hiking with a steaming towel draped over my head. The rain finally eased, but that just made things hotter, as heavy sunlight hammered down through the parting clouds. Slipping in the mud

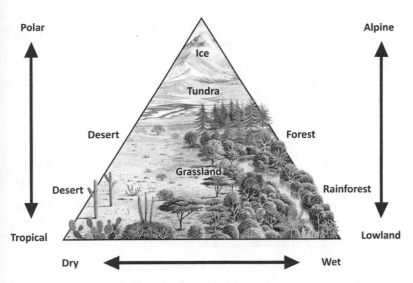

FIGURE 11.3. Leslie Holdridge's original life zone diagram was all text, numbers, and lines, but this illustrated rendering captures its essence, showing how temperature and moisture interact to define habitats across the globe. Illustration © Chris Shields.

and tripping over slick tree roots, I had to earn every foot of elevation gain the hard way, and soon realized that I wasn't going to get nearly as far as I'd planned. But even so, I could already see changes in the vegetation around me. The giant almendro trees I'd been studying—so common in the lowlands—had disappeared completely, and the stature of all the canopy species seemed shorter. I noticed more ferns in the understory and fewer big palms, and while the vegetation was still dense, there were also occasional gaps that opened up views out over La Selva and beyond. In terms of Holdridge's triangle, I was in the lower right-hand corner, ascending from wet rainforest toward something he called the "pre-montane transition." With more time (or a helicopter), I could have continued on up into the cloud forest and ended in a high montane woodland of moss-covered oaks. And that's just Volcan Barva. Costa Rica's higher peaks reach up into the tundra-like grass zone known as paramo, and the country's far northwest lies in a seasonally parched rain shadow, with scrub forests dry enough for acacias and cacti. By one recent count, Costa Rica boasts twenty-three distinct life zones—nearly two-thirds the number found in the entire continental United States, all packed into an area about the size of Denmark.

For Leslie Holdridge, the concentrated diversity of climates and habitats in Costa Rica provided an ideal, perhaps irresistible, proving ground for his theory. He first introduced the life zones concept in 1947, publishing a three-page outline that ended with the sentence, "Further details and examples will be given in a paper now in preparation." That preparation ended up lasting for nearly two decades, as Holdridge painstakingly field-tested his ideas throughout Costa Rica's many life zones and beyond. By the time the resulting two hundred–page opus was published, it was almost beside the point. Scientists had been citing its three-page predecessor for years, applying the life zone concept to everything from bird and frog distributions to the geography of Peru. In short, Leslie Holdridge's theory was becoming part of the scientific firmament, and as such it would go on to be adopted by (and adapted by) a wide range of disciplines. Even when the resulting models veered widely from his original design, they all shared his vision of habitat as an abstract, multidimensional space, something that could be defined and manipulated by adjusting certain variables. The term *envelope* soon came into use for this concept, borrowed from the way pilots balance the effects of air speed, load, and lift to calculate a safe flight space. And just like flying an airplane, there was the implication that pushing habitat variables too far outside of their envelope could bring the whole system crashing down.

When Leslie Holdridge first conceived of life zones, the idea that carbon emissions were altering the atmosphere was, to use his word, "conjecture." He saw plant communities as relatively stable, and he considered his system descriptive rather than predictive. But by the 1980s, researchers concerned about global warming had embraced life zones as a fundamental method for envisioning the future. Whenever new climate models produced better temperature and rainfall projections, those numbers could simply be plugged into the Holdridge system. Make conditions hotter, and outcomes shifted downward on his triangle; make them drier, and everything moved to the left. One prominent early climate paper invited readers to do just that—it included a full-page reproduction of the Holdridge diagram, along with predictions about forests and shrublands rapidly transitioning into arid grasslands and deserts. As the field of study has advanced, climate and biological forecasts have gotten far more sophisticated. Researchers now parse the climate into dozens of different variables, zeroing in on which nuances matter most, not just for communities and habitats, but for the individual species that call them home. Is average temperature more important than extremes? Are certain seasons more relevant? What about storms, floods, droughts, or nonclimatic factors like soil type and terrain? For scientists, the result is a dizzying array of potential model designs, described by an equally dizzying array of acronyms: GLMs (generalized linear models), GAMs (generalized additive models), PRISMs (parameter-elevation relationships on independent slopes models), or CEMs (climate envelope models), just to name a few. The challenge lies in finding mathematical expression for the climate-driven patterns so readily observed in nature. That was Holdridge's driving passion, and that's why—even though he never studied climate change directly—biographers now describe him as a forester *and* a climatologist.

I can't say that I noticed the heat dropping off in my brief uphill scurry from La Selva, and I doubt that my socks smelled any better by the end of it. But the vegetation was obviously re-sponding to the change, even in that short vertical distance, and if half a century of biological modeling has taught us anything, it's that small differences matter. The sophisticated computer simu-lations, data mining, and other methods required to understand those nuances can seem daunting (even to biologists), but the re-sults don't have to be. One of the most ambitious projects to date can be fully explored online, and—for anyone living in North America—field-tested right in your own backyard.

Rigorous biology, like all good science, demands objective, clear-eyed thinking devoid of sentimentality. But the fact is that every-one has a few favorite species. Personally, I've always been partial to the golden-crowned kinglet, a pixie-sized songbird that seems to embody the damp forests where I live—as gray and green as the clouds and the trees, and with a fiery crest that rises up only occa-sionally, like sunshine breaking through on an overcast day. I re-member kinglets from childhood, so common they would gather at our feet to drink from the hose drips while we watered the gar-den. In graduate school, I spent a few weeks studying their winter flocking behavior—being so reliably easy to find made them the perfect subjects for a short-term research project. Only recently did it occur to me that their flitting presence was getting scarce in the woods around our house. Somehow, my favorite backyard bird seemed to be on the wane.

You have to look pretty hard to find words like *guesstimate* or *hunch* in a scientific journal, because researchers always use hard data to support or refute their hypotheses. Without good numbers on the kinglet population, I couldn't know whether the decline in my neighborhood was real or if it was an artifact of some other trend—middle-aged eyes less able to spot tiny birds, for instance, or ears growing deafer to their chimelike calls. For a quick data fix, I turned to one of the largest repositories of bird information anywhere, the records of the National Audubon Society's Christmas Bird Count. Founded in 1900 as a benign alternative to the holiday shooting sprees popular at the time, the annual volunteer tally has grown from 25 original locations to more than 2,500, spread across North America, South America, and beyond. People in the islands where I live began contributing in 1985, prowling the woods and beaches on a chosen day each winter, and reporting every bird they encountered. Not surprisingly, golden-crowned kinglets have shown up on every annual list. But when I graphed the data there was no question about where their numbers were headed. In the past five years, local sightings had dropped over 65 percent from their 1985 to 2000 average, and some counts were even worse. Scores of birdwatchers working together in 2017 on a fine, sunny day hardly managed to spot one kinglet for every hour of searching.

FIGURE 11.4. A golden-crowned kinglet at home in a fir tree. Modeling studies can help predict where (or whether) this will still be a common sight in the future. Depositphotos.

Now that I knew I wasn't imagining things, I could turn my attention to what might be causing the trend. With so many climate-driven range shifts under way, it seemed likely that such a cold-hardy species might be moving northward. But testing that theory would require a lot more data than a few annual observations from one location. Luckily, the folks at Audubon do more than count birds at Christmastime. Their scientific staff had recently completed a detailed habitat model that showed precisely where climate change was driving golden-crowned kinglets, right down to the zip code. And while they were at it, they did the same thing for 603 other North American birds.

"We really wanted to understand the impacts for as many species as we could," Chad Wilsey told me. As Audubon's chief scientist, he oversaw a massive effort to gather and process more than 140 million site-specific bird sightings, looking beyond the Christmas Bird Count to draw on university researchers, government agencies, and thousands of individual birdwatchers who contribute to the online eBird platform. It's no wonder Wilsey's team had to partner with a data science company to manage the analysis. Even with the power of cloud computing, it took months just to finish the initial stage of the project: determining where the birds live now, and, perhaps even more crucially, *why* they live there. Detailed maps of all those millions of observations satisfied the first part of the question, but answering the second part required an approach that's relatively new in ornithology—the use of artificial intelligence (AI).

"Machine learning is a powerful tool for extracting patterns from voluminous data," Wilsey explained to me over the phone, and summarized the process as if it were as simple as blending up a smoothie or a milkshake. "Observations go in, models are tested, and the best fit comes out." For Wilsey, that level of comfort came from long practice—he'd been using AI-based techniques in his bird research for well over a decade. Like a latter-day Leslie Holdridge, Wilsey got his start modeling habitats in Costa Rica. "That's where my passion for birds and conservation was born," he told me, and said he had always been focused on helping birds and people coexist in the same landscapes. "I'm interested in how models can inform better management," he explained, and indeed his work has contributed to habitat management decisions in a range of contexts, including military bases and sites being developed for natural gas extraction. Wilsey spoke with the self-assurance of someone accustomed to explaining abstractions, and it was tempting to skip straight ahead to the results—where the kinglets and other birds were likely to move. But I wanted to understand the nuts and bolts of the process, so I pressed him to explain the middle part, the step where "models are tested." What happened after he and his colleagues entered all that voluminous bird data and pushed the button? What was the computer actually doing?

There was a pause on the line, and I could hear the sound of a clicking keyboard in the background and the ring tone from another phone. (It occurred to me that, much like the computer models he specialized in, Chad Wilsey probably had a mind capable of doing many things at once.) "Think of the model as a repeatable process that leads to learning," he said at last. "That's really the definition of an algorithm. My favorite example, and the one I'm most familiar with, is called Random Forest." He went on to describe how Random Forest takes small subsamples of data and builds simple models, called decision trees, to explain them. For any particular species, those decision trees use climate variables and other factors to explain where the birds were seen. Was the average spring temperature above x? If yes, was the annual rainfall less than y? . . . and so on. By repeating that process thousands of times, using different subsets of data and different combinations of questions, the algorithm creates a "forest" of potential models. Some models are clearly better than others at explaining the data, and the successful ones help reveal which variables are most important for the species in question. "You experiment with it," Wilsey said. "You let the tool separate the signal from the noise."

What emerged from the Audubon analysis were models that identified both summer and winter range requirements for every species, defined by the relative importance of more than a dozen climate and habitat variables. Certain birds responded to changes in temperature more than rainfall, for example, or to things like the number of frost-free days, the ruggedness of the terrain, or the presence of wetlands. Once those "best fit" models had been built and vetted for each species, making predictions became a straightforward exercise in mapping. If kinglets, for example, lived only in forested habitats within a certain temperature and moisture range, then it was simply a matter of identifying where those places would be when the world is warmer. Standard climate change projections provided those answers under a variety of future climate scenarios, and Audubon published the results. But in addition to the expected peer-reviewed papers and reports, they took the time to produce a flashy, interactive website. Seeing those maps in full color captures what may be the most crucial insight that modeling brings to climate change biology—a glimpse of where species will have to go to find the conditions they need. For Wilsey and his team, however, there was another question that was even more important: What chance do the birds have of getting there?

"I was surprised at how well the models fit our predictions of vulnerability," Wilsey said, and explained that a second layer of analysis helped them identify which birds faced the steepest challenges in the years ahead. Maps of habitat lost versus habitat gained, combined with measures of adaptability (like how far young birds tended to disperse from the nest), produced what amounted to a risk assessment for each species—just the sort of management-ready tool that Wilsey always aimed for. "We wanted to make the results as useful and relatable as possible," he confirmed, and not just to managers and scientists. "The public was very much in our minds as a key audience," he added. With that in mind, I went online immediately after Chad and I said our goodbyes and began exploring the future of the golden-crowned kinglet.

The first thing I noticed was the map's color scheme. A dramatic splash of dark red represented habitat loss all across the southern portion of the kinglet's range. Even the most optimistic level of warming showed a red patch on the island where I live. (Fittingly, it seemed to be the pixel right on top of my house.) But as I clicked through the various charts and projections, I saw that there were also green places that marked improved habitat, and pale blue ones where the kinglet could expand into new territory. As suspected, those opportunities all lay to the north, and if my backyard observations were any indication, the kinglets were already racing to get there. At least they had a place to go, and Wilsey's team ranked them as only moderately vulnerable to climate change. The situation brought to mind something marine biologist Gretta Pecl had told me—if a species can move and survive, that's a great outcome. So I logged off my computer feeling somewhat encouraged about my favorite little backyard birds. I might miss them, but I shouldn't mourn them. They were just moving somewhere else. Making such distinctions is vital, because the resources to help struggling species are limited—not just in terms of scientific and conservation capital, but in emotional capital too. With the world changing so rapidly, biological modeling provides something personal as well as something practical: it helps us decide what to worry about.

By chance, I spoke with Chad Wilsey on the day his team published a follow-up paper to their modeling project, one that put its predictive powers to the test. Teams of professional ornithologists had partnered with volunteers across the country to search for particular birds in particular landscapes. They found that the abundance of their target species—various nuthatches and bluebirds—were indeed living in the places predicted by their climate models. Habitats determined to be optimal held significantly more birds, while populations were diminished in marginal areas. But perhaps the most striking result had to do with colonization: seven examples of birds that were expanding their ranges into areas identified as newly suitable—those promising blue patches on the range maps. Wilsey admitted that the validation was gratifying, but he cautioned against taking any model prediction too literally. "There is always uncertainty," he said, and reminded me of the difference between correlation and causation. Models can identify useful and telling patterns without necessarily revealing what makes them happen. Kinglets, for example, may indeed be heading north to find the cooler temperatures they prefer, just as predicted, but no one knows precisely why. It would take a lot more research to figure out just what it is that the birds don't like about warm weather.

Similar to the Audubon study, most other climate-related models in biology have focused on species distribution, identifying places where plants and animals are likely to occur (or not) in the future. But, as we know from earlier chapters of this book, rising temperatures are not the only challenge brought on by climate change, and movement is far from the only response. Many potentially important variables don't lend themselves as easily to modeling—things like unexpected plasticity or rapid evolution, for example, or changes to crucial relationships like predation, pollination, and parasitism. There is a point where any model simply can't include all the complexities of life in the real world. As an alternative, some biologists have begun transplanting species, seeing how they fare in warmer places that resemble the conditions they'll soon be facing at home. Alpine plants and pollinators have been shifted to lower elevations, for example, and corals have been moved among reefs with varying water temperatures. These experiments are also limited, however, since it's impossible to move entire communities and all the intricate interactions that sustain them. To overcome that challenge, scientists working in the bogs of northern Minnesota have come up with a creative and elaborate solution. They're changing the climate of their study site.

"We call it a whole ecosystem warming experiment," Randy Kolka told me, just before his face froze. With rural internet connections on both ends, our Skype call was lucky to suffer only the one interruption, and after a few minutes of clicking and waiting, Kolka's home office came back into view. Webcam backdrops had become something of a statement during the ongoing coronavirus lockdown, and I noted the trophy-sized muskie, a large species of pike, mounted on the wall behind him. "I've been stuck in a corner of my basement for three months," he groused, and for a moment I wondered which he regretted more—the impacts to his research, or the missed fishing opportunities. For my part, the stay-at-home orders had meant canceling a trip to visit Kolka and tour the field station known as SPRUCE, short for Spruce and Peatland Responses Under Changing Environments. It's a long name, but as Kolka pointed out, it's also the largest climate manipulation experiment on the planet.

"It's absolutely unique," Kolka assured me, and from the pictures he sent I had to agree. Looking more like science fiction than science, the project site rises up from a flat, forested bog in

FIGURE 11.5. Randy Kolka on the boardwalk at SPRUCE, where huge open-topped terrariums mimic the soil temperature, air temperature, and carbon dioxide concentration of various climate futures. Photo © Layne Kennedy.

a series of ten octagonal pods, each one over two stories tall and fashioned from shining panels of glass and steel. The open-topped chambers enclose thousands of square feet of habitat, complete with standing trees and shrubs, where the air temperature can be manipulated, and where carbon dioxide gets pumped in around the clock to simulate a range of atmospheric futures. But with a multimillion-dollar budget and over one hundred collaborating investigators, SPRUCE had the resources to go even further—inventing and installing an ingenious system of buried pipes to heat the very ground itself. As a soil scientist for the US Forest Service, Kolka helped spearhead that aspect of the project. It was an unusual step, but one that added an extra dose of realism to the experiment, shining a research spotlight on a hidden realm too often overlooked by climate change biologists.

"We're already seeing changes in elevation," he said, and explained how the added heat was boosting activity underground, improving conditions for many of the microbes and other soil dwellers that regulate cycles of rot and renewal. The upshot was an increase in decomposition, measurably lowering the level of the bog as the peat layer began to dwindle. For climate scientists, that result alone made the whole project worthwhile, because the rate of decay in bogs is pivotal to determining the rate of global warming. "It's all about carbon," Kolka told me, and described how the plant remains that form peat accumulate over millennia in soggy, acidic conditions. At the SPRUCE site, those deposits reached more than ten feet (three meters) in depth and dated back eleven thousand years. "Peatlands take up only 3 percent of the Earth's land surface, but account for 30 percent of total soil carbon," he went on. In other words, peatlands currently act as carbon "sinks," removing carbon from the atmosphere and storing it long-term underground. But if a warmer future brings faster decomposition, then those deposits will begin to rot away, and all of that stored carbon could be released. "They could flip," Kolka told me. "Peatlands could turn from carbon sinks into carbon sources."

Many questions remain about when, where, and at what temperature that tipping point might be reached, and whether or not it will penetrate deep enough to impact the truly ancient peat. But SPRUCE gives scientists like Kolka a chance to move beyond algorithms and models, testing their predictions with real temperatures on living microbes in a real bog. As an added bonus, they get to see how the rest of the site's inhabitants respond. "There are huge changes going on," he said, and explained how higher temperatures were extending the growing season and making the site drier, triggering a number of expected trends like the expansion of woody plants and the decline of *Sphagnum* moss. New species were also arriving, taking advantage of the altered conditions, and the added carbon dioxide was having an effect too, giving many plants at least a temporary boost in growth rates. But as we talked, what caught my attention were the observations and findings that, in Kolka's words, "were not in our original hypotheses."

"The trees in the warmest chamber are hating it," he said, an apt description of how the project's namesake, black spruce, had unexpectedly begun to decline. Although the additional heat did make the growing season longer, it also created more "false springs" in late February and March—abnormally warm days that tricked the trees into breaking bud early, only to see all that fresh growth killed off during the next cold snap. "They only have enough reserves to do that once in a while," Kolka said. If it happens too often, the trees wither and eventually die from what amounts to botanical exhaustion. Nobody saw that result coming, and it's still a mystery why other woody species don't suffer the same fate. In fact, certain shrubs in the hottest chambers are positively thriving—spreading rapidly, getting larger, and producing bigger, juicier fruits. "If you like blueberries, the future looks good," Kolka quipped.

With studies in progress on everything from lichens to sedges to spiders, SPRUCE will no doubt continue to produce a few surprises alongside the predicted results. Kolka and his colleagues are already proposing new experiments and testing new hypotheses, hoping to extend the project beyond its proposed ten-year lifespan. That's how science works—each discovery leads to additional questions, even when (or especially when) the results are unforeseen. For climate change biologists, the element of surprise is turning out to be a very familiar part of the job description. Because, as the next chapter shows, even seemingly straightforward relationships and predictions can veer off in unexpected directions.

CHAPTER TWELVE

Surprise, Surprise

Our experience shows that not everything that is observable and measurable is predictable, no matter how complete our past observations may have been.

—Sir William McCrea
"Cosmology—A Brief Review" (1963)

The most famous butterfly in chaos theory started out life as a seagull. Meteorologist Edward Lorenz introduced the analogy in a 1963 lecture about the limits of predictability, suggesting that changes in the atmosphere as small as the flap of a gull's wing could produce ripple effects with large, unknowable consequences. On the advice of a colleague, he later replaced the bird with a colorful insect, and "the butterfly effect" was born. Lorenz meant it as a comment on the difficulty of predicting complex systems, though it's also invoked to suggest how small changes can have unforeseen consequences. Either way, it turns out to be a good analogy for the biology of climate change.

In a nutshell, chaos theory looks for order within disorder, the underlying patterns hidden in randomness. It's no surprise to climate watchers that it arose from the study of weather. By one read, Lorenz's storied career never strayed far from the central frustration of his profession: why it's so difficult to provide an accurate, long-range weather forecast. Biologists face a similar challenge when trying to predict the results of climate change. Certainly, there are a few agreed-upon expectations: many species will move; some will adapt, and some will be lost; new communities will form; and flexible generalists will have a great advantage over specialists. But natural systems are every bit as complex as the weather, filled with wings that flap literally as well as metaphorically. The potential for butterfly effects is enormous, which gives biological oddsmakers at least one prediction that can be made with confidence: expect the unexpected.

Jane Austen once observed that "surprizes are foolish things. The pleasure is not enhanced, and the inconvenience is often considerable." Some scientists might agree, at least with the latter point. It's always disheartening (and often expensive) to watch a carefully planned experiment or field season suddenly thrown into confusion by something unforeseen. But inconvenient does not necessarily mean unproductive, and surprises in science often lead to important new discoveries and ideas. We've already encountered a number of them within these pages, from diet-shifting bears to misplaced pelicans to lizards evolving almost overnight. But those cases can all be largely explained by the unanticipated speed of climate-driven change. Biologists went into the field anticipating something conventional, only to be surprised by a new set of conditions. It's another matter altogether when climate change models themselves go awry, when some overlooked detail upends a thoughtful prediction and sends the real-life outcome spinning off in a different direction. That kind of surprise is happening more and more, putting theory to the test wherever climate realities are catching up with expectations. Even projections that seem simple and ironclad can go off the rails, like the case of a common seabird in a warming Arctic wilderness.

Franz Josef Land is an archipelago that lies within Russia's Arctic National Park. It contains the northernmost outcrop of dry land in Eurasia and is located barely 550 miles (900 kilometers) from the North Pole. Sea ice surrounds the islands for much of the year, providing ample foraging opportunities for polar bears, walrus, bearded seals, and other ice-dependent creatures. Among the smallest on that list is a black and white seabird so endearingly plump and round it looks like a plush doll come to life. The dovekie, or little auk, outnumbers every other seabird in the far north, its abundance memorably described by nineteenth-century explorer Captain Frederick Beechey: "They are so numerous, that we have frequently seen an uninterrupted line of them extending full half way over the bay, or to a distance of more than three miles, and so close together that thirty have fallen at one shot. This living column, on an average, might have been about six yards broad and as many deep; so that, allowing sixteen birds to a cubic yard, there must have been nearly four millions of birds on the wing at one time."

The flock that Beechey described was leaving a cliffside nesting colony, heading seaward to their preferred feeding grounds along the edge of the pack ice. Like other auks, dovekies are diving birds, using their stubby wings to "fly" underwater in pursuit of prey. But where larger members of their tribe mainly pursue fish, dovekies focus on zooplankton, tiny crustaceans that thrive in great swarms where meltwater mingles with the cold brine of the Arctic Ocean. That affinity for ice margins makes them particularly vulnerable to climate change. In Beechey's day, the frozen seas rarely retreated more than a few miles from shore in places like the islands of Franz Josef Land, which meant that those great flocks he witnessed could reach their feeding grounds easily. Now, the ice margin is shifting farther north with every passing year, presumably making it more and more difficult for breeding dovekies to feed themselves and provision their growing chicks. With summer sea ice in the Arctic expected to disappear altogether as early as 2050, predictions for the dovekie population were fairly straightforward: steady decline, followed by a crash. Yet when ornithologists went to the field to test that model, they found that a very different version of the future had taken hold.

"The whole trip to Franz Josef Land was a surprise, and an incredible adventure," David Grémillet wrote to me in an email. Currently the French National Centre for Scientific Research (CNRS) senior scientist and director of the Chizé Center for Biological Studies at La Rochelle University, Grémillet participated in a major expedition to the islands in 2013. It was a joint exploration by Western and Russian scientists, involving dozens of researchers investigating everything from algae to geology to marine viruses. Grémillet's group spent nearly a month at an abandoned Soviet station in Tikhaya Bay, which he described as "like a great open-air museum of USSR in the 1950s, with all its wooden barracks left untouched and slowly filling with ice." The site allowed them easy access to a nearby breeding colony of dovekies that numbered in the tens of thousands. Having already studied the species in Greenland and Norway, Grémillet's team followed a well-established protocol. "We proceeded as we always do," he wrote, "by catching the little auks near their nests, and fitting them with 3-gram electronic trackers." The surprises began when they recaptured the birds, relieved them of their tiny trackers, and started downloading the data.

"We had indeed strong hypotheses and predictions on how they would behave," he explained, noting that birds in their previous studies had regularly flown over sixty miles (one hundred kilometers) to reach the edge of the pack ice. "We were expecting flight times between the colony and the foraging spot of at least an hour," he went on, and then recounted what he called "one of the most exciting moments of my research career." Sitting with their laptops at the dinner table, surrounded by their Russian counterparts, they opened up the first batch of tracking data and saw precisely how long the birds had been in the air: less than four minutes. Instead of trekking all the way to the edge of the sea ice, the dovekies had apparently found an alternative food source right on their doorstep. But what, and where? It's easy to imagine the excited conjecture and conversation that came next, perhaps fortified with a sip or two of vodka. Soon their ideas began co-alescing around an entirely new hypothesis.

"My colleague, Jérôme Fort, remembered what we had seen while climbing the nearby mountain with our Russian friends a week before," Grémillet recalled, and described a distinct line across the mouth of the fjord, where cloudy blue meltwater from island glaciers slammed into the dark, dense currents of the Arctic Ocean. Both Fort and Grémillet had trained as oceanographers before they began studying birds, so they understood the consequences of such an abrupt transition. "We both knew what this front meant: a curtain of plankton, killed by temperature and osmotic shock." For tiny crustaceans, swimming so suddenly from one kind of water into another was like driving full speed into a wall. And for anything that fed upon those crustaceans, the resulting pileup created a bonanza.

Testing their theory required a boat, but the only craft available was "a chronically deflated dinghy," and the fuel they'd brought from Murmansk turned out to be contaminated with water. Not exactly ideal for exploring a polar sea, but they set off anyway and managed to sputter far enough out into the fjord for a survey. There wasn't much to see at first, but as they crossed

FIGURE 12.1. The dovekie, or little auk, has defied climate change predictions by taking advantage of new foraging opportunities brought on by melting Arctic glaciers. Photo © David Grémillet.

the convergence between glacial melt and ocean water, the dovekies were suddenly all around them. "The little auks were all there," Grémillet wrote, "aligned on the oceanic side . . . diving and stuffing themselves on plankton, easily picked from an underwater curtain."

With that revelation, the story of dovekies and climate change switched instantly from one of decline to one of resilience. Yes, the sea ice was melting as predicted, but so were Arctic glaciers. And in places like Franz Josef Land, where glaciers are plentiful, that created an opportunity that no one saw coming. Grémillet and his team spent the remainder of their field season showing that dovekies weren't just surviving on their new food source, they were thriving. Chicks grew at precisely the same rate as they had on a traditional diet, measured decades earlier at the same location. The only possible signs of stress were in the adults— greater variation in their diving behavior and a slight decline in body weight, as if feeding at the "underwater curtain" might require a little more effort. To Grémillet, the project demonstrated how one overlooked detail can have a huge effect on the outcome. "Even if you think that you know," he summarized, "you really have to get out into the field to check what wild creatures are doing, because they very often surprise you."

At current rates of warming and melting, the glaciers of Franz Josef Land should hold out (and keep making plankton curtains) for up to 180 years. After that, it's anyone's guess what the local dovekies will do. Ultimately, the most important finding from Grémillet's study may not be what the dovekies were eating or where, but how easily they made the switch. (In other words, their plasticity.) Rapid dietary changes have also occurred for dovekies in Greenland, where the birds now feed on the larvae of mackerel and other newly arrived warm-water fishes. But Grémillet still considers the species highly vulnerable, because, as he put it, they are living on "an energetics knife edge"—struggling to find enough calories year-round in an environment that was harsh to begin with. And if one of his latest ideas is correct, there are changes coming that may upend dovekie predictions even further. Almost as an afterthought to our email exchange, he forwarded a brandnew paper with "potentially spectacular" implications. If summer sea ice does disappear, what's to stop North Atlantic seabirds from migrating over the pole? He and his colleagues calculated that dovekies (and potentially many other species) would save considerable energy by spending their winters in the warmer North Pacific, and could potentially establish breeding colonies there as well. For future ornithologists, the resulting geographic confusion might indeed seem like chaos. It's one thing if you don't find your birds feeding in exactly the way you predicted; it's quite another if you're not even looking for them in the right ocean.

When changing the inputs of a system or equation doesn't predictably change the outputs, mathematicians call the relationship nonlinear. Biologists use that word too, particularly when describing unforeseen connections like the one between dovekies and glaciers. In practice, using *nonlinear* often amounts to a coded admission of surprise at one's research results, and the term is becoming increasingly common in the biological literature as more and more climate predictions are put to the test. The impact of false springs on the SPRUCE project's bog trees is a good example, and botanists have found similar effects farther north, where swathes of tundra and boreal forest are suffering the counterintuitive fate of death by freezing. (Though the plants benefit from a longer, warmer growing season, reduced snowpack in winter exposes them to lethal cold.) One of the most common climate responses—early springtime flowering—is also easily disrupted, or even reversed, by changes in rainfall, by temperature changes during other seasons, or by site-specific factors like elevation and aspect. Pollinators can make the situation even more unpredictable. When bumblebees don't find enough spring flowers to feed on, for example, they sometimes start chewing holes in leaves to spur their favorite plants into blossom. The stress of that physical damage appears to trigger a "now or never" reproductive surge, advancing the flowering period by as much as a month, regardless of the weather. It's fascinating biology, but highly local and erratic—tough work for a predictive algorithm.

Connections that are overlooked, or simply unknown, will always guarantee a certain element of surprise in climate change biology. But there is another chaotic ingredient in the mix that is even more reminiscent of the butterfly effect—the unforeseen consequences of events that take place at a great distance. One of the best examples takes us right back to the landscape where this book began, Joshua Tree National Park, and to the signature species that gives the park its name. The story is biological, but it also involves paleontology, because the key event that explains it wasn't distant from the park in space, it was distant in time.

"Year after year went by and I could never get it funded," Ken Cole lamented, remembering the long struggle to get his Joshua tree project off the ground. He'd noticed a distinct pattern in the field—adult trees were dying off with few offspring to replace them, and what seedlings he could find seldom grew more than a hundred feet (thirty meters) from their parents. He had a solid hypothesis to explain the situation, and when he pitched the project to the US National Park Service, they thought it was a great idea. So did his bosses at the US Geological Survey, where he'd already established himself as an expert on climate and plant relationships in America's desert southwest. Yet, somehow, the fate of one of the region's most iconic species never quite reached the top of either agency's priority list. Ken and a colleague kept gathering preliminary data to bolster their case, making up for the lack of funds with sheer ingenuity.

"The hardest part was figuring out current distribution— where exactly do Joshua trees grow?" Cole recalled. For such a famous species, surprisingly little information was available, and he couldn't afford to hire a field crew to go out and conduct surveys. But Cole refused to give up, and he eventually hit on a solution that wasn't just low cost, it was no cost: real estate ads. By searching through online property listings, Cole began amassing a huge dataset about potential Joshua tree habitat. Every time a house or a plot of land came up for sale in and around the Mojave Desert, he got a new data point with a precise address. And since the listings always featured photographs, all he had to do was scan through the picture gallery and look for a familiar shape. "It was pretty easy to tell if there were Joshua trees there."

After eight years of doing research on the cheap, Cole realized that he had enough data to analyze and publish his findings, without ever having received a grant. It helped that his colleague was a whiz with species distribution models, and it didn't hurt that Cole already had what, for most people, would have been the harder information to gather. He knew the history of Joshua trees, all the places where they'd been growing—or not—since the Pleistocene. It was a dataset that spanned more than thirty thousand years, and he'd assembled it by studying a small desert dweller that most people dismissed as a pest.

"Pack rat middens are loaded with Joshua tree fossils," Cole explained, and told me how he'd devoted his doctoral research and much of his early career to picking through old rat nests, identifying plant remains. That might not sound very glamorous, and for the nests of the average rodent, it wouldn't be. But pack rats don't just tuck away a few bits of fluff and a cache of food; they are compulsive hoarders, building eclectic and sometimes enormous piles of debris that can include everything from leaves and seeds to bones, insect remains, and shiny buttons. Those treasures are all gathered within a few hundred feet of the cave or rocky crevice they call home, so each pack rat midden provides a snapshot of its immediate surroundings. That matters to scientists like Cole, because in the desert those middens can persist indefinitely, preserved by a combination of dry air and the way pack rat urine crystallizes to form a hard, amber-like crust. In a sense, analyzing the data from pack rats isn't so different from looking at real estate ads—you have a location, and you get a glimpse of the local vegetation. Putting the ads and the rat data together provided Cole with a long-term geographic history of the species, telling him how the range of Joshua trees had expanded and contracted across the landscape in response to past climate trends. And it also told him something else. This time around, something was different.

"Joshua trees are ideal for this kind of study because they respond directly to temperature," Cole said. That trait made it easy for him to map and model their historic range—retreating south into modern-day Mexico during cool periods, and shifting northward when the climate warmed. But that long-standing pattern apparently went right out the window after the end of the last ice age. Yes, populations to the south began dying out as the world warmed up, a trend now accelerating with modern climate change. But for some reason the other half of the equation—northward expansion—had utterly ground to a halt. A large belt of land in southern Nevada and adjacent California had become ideally suited to Joshua trees, but the plants showed no signs of moving in. And Ken Cole has a pretty good idea of why.

"A lot of things in my career came together in this project," Cole mused. Though technically retired now, and sporting a fluffy white beard, he told me that he still works pretty much every day, and just before our video call he'd hiked several miles into the desert to retrieve a remote weather monitor. But his key insight about Joshua trees dates back decades, to a field trip he took in graduate school. "We were in a cave in the Grand Canyon where giant ground sloth dung was piled ten feet thick," he said, and recalled how his doctoral advisor had pointed to the top of the pile and made a dramatic proclamation: *"That's the last dung ball that fell there before giant ground sloths went extinct, twelve thousand years ago."* Cole remembers the moment vividly not just for the setting and the statement, but for what those ancient droppings obviously contained—visible fragments of Joshua tree leaves and fruits, just like the pack rat middens he was working on.

To some people, this might have seemed like nothing more than an interesting coincidence—that one of the ancient desert's largest inhabitants left behind the same sorts of plant remains as one of its smallest. But for Cole the implications were enormous. "It got me thinking about why Joshua tree fruits were so high off the ground," he said, and then listed a number of other traits that made the fruits unusual for a yucca. "They don't break open; they're the size of a lemon; they're fleshy; they're nutritious." In other words, where most yucca plants sported low, podlike capsules that split open to disperse their seeds, Joshua trees stood tall, with succulent fruits obviously intended to attract animals. Yet what animals? It's true that pack rats would gnaw them open to get at the seeds, but only when they found the fruits on the ground, old and dried out. It seemed odd that nothing ate them at the peak of ripeness, when they boasted a sugar content as high as 25 percent and clustered in prominent green displays, weighing down the branch tips. In the words of one researcher, why would Joshua trees "expend large amounts of energy and resources to create a product for which there is no market"? The answer, of course, was that the market had disappeared.

"Shasta giant ground sloths died out at the same time as mammoths and other megafauna," Cole said. He chalked up their extinction to something called Pleistocene Overkill, the idea that when human hunters migrated from Asia to North America at the end of the last ice age, their efficiency with spears and other weapons rapidly eliminated dozens of large mammal species from the continent. (Not coincidentally, Cole's doctoral advisor—the one who first brought ground sloth dung to his attention—was none other than the theory's main architect, Paul S. Martin.) But regardless of why the sloths perished, their absence continues to reverberate biologically.

Measuring up to nine feet in length and weighing in excess of 550 pounds (250 kilograms), adult Shasta giant ground sloths were perfectly sized to feed on Joshua tree fruits, and fossil evidence from their dung heaps suggests that they did so regularly, not just in the Grand Canyon, but all across the desert southwest. It was an ancient partnership that provided sloths with calories and gave Joshua trees a reliable long-distance seed disperser, allowing their range to expand and contract in rhythm with varying climate patterns. By mapping where Joshua trees grew in the past and modeling where they should have moved since the climate began to warm, Cole's study showed how their dispersal had stopped precisely when—to borrow Paul S. Martin's memorable words—the last dung ball fell. Without those giant creatures ambling for miles across the landscape, Joshua trees must now rely solely on pack rats and other scurrying rodents to move their seeds, a strategy that yields a paltry dispersal rate of only six feet (two meters) per year. With climate change rendering more and more of their current range inhospitable, Joshua trees find themselves haunted and hampered by the past—unable to reach cooler conditions to the north, and at risk of disappearing from, among other places, the national park that bears their name.

Before the end of our conversation I gave Ken Cole a thought puzzle. If Shasta giant ground sloths had survived and were still roaming the American southwest, would Joshua trees be keeping pace with modern climate change? He smiled and paused, but only for a second, as if complex calculations of warming and dispersal traveled a well-worn path in his mind. "Yes," he concluded, "they probably would be." Instead, Joshua trees may soon be confined to north-facing slopes and other cool refugia, or, ironically, they may have to rely upon us humans to be their new long-distance dispersers. "People already like growing them

FIGURE 12.2. When the Shasta giant ground sloth went extinct, Joshua trees lost their main long-distance seed disperser, an impact being felt now, thousands of years later, as they struggle to keep up with a changing climate. Illustration © Chris Shields.

in gardens," Cole observed, so it would only be a matter of scaling that up to establish viable populations in northern locations. Biologists call that strategy human-assisted migration, and it's under consideration for a growing number of species, because Joshua trees are not alone in finding their ability to move unexpectedly confounded or blocked. Nor are ancient extinctions the only phenomena interfering with natural responses to climate change. Habitat loss, urbanization, pollution, invasive species, and other human-driven trends have drastically altered ecosystems, muddling countless evolutionary relationships and strategies in the process. Plants and animals now find themselves facing climate challenges in an environment already quite different from the one they evolved in and adapted to. That scenario adds yet another element of chaos to the business of making predictions, further ensuring that climate change biologists will always have a lot of surprises to look forward to.

In his 1952 story, "A Sound of Thunder," science fiction author Ray Bradbury presaged the idea of the butterfly effect, quite literally. He wrote of time travelers on a trip to the Jurassic period who inadvertently crush a single gold and black and green butterfly underfoot. Returning to the present, they find their world subtly but profoundly altered. Words are spelled differently, people talk strangely, and the results of a recent presidential election are reversed—all consequences of that one tiny change rippling forward through the ages. Biologists equipped with a time machine would also be interested in visiting the past; not to change it, but to learn from it. On a warming planet, the best guide to the future may well lie in what has gone before, because while the drivers of this episode may be different, any examination of history makes one thing clear: climate change is nothing new.

CHAPTER THIRTEEN

That Was Then, This Is Now

The historian is a prophet facing backwards.

—Friedrich Schlegel
Lyceum (1798)

The saying "You can't go home again" is particularly true if someone else now owns the house. Fortunately, it wasn't the residence of my childhood that I wanted to revisit. I was interested in seeing the alley behind it, a public right-of-way that stretched from the back wall of our garage six blocks south to the nearest major cross street. As I parked the car and got out to look around, the place matched up fairly well with my memories—a narrow, paved lane cluttered with garbage cans, lawn furniture, boats on trailers, and a range of other domestic accumulations. Behind one home, I was pleased to see a swing set and a scattering of toys, and other yards held basketball hoops, kiddie pools, and at least two homemade skateboard ramps. When I saw a hand-painted sign that read "Slow—Children at Play," I knew that one thing about the neighborhood clearly hadn't changed. Different families might occupy the houses, but the alley still belonged to the kids. In the years that I lived there, saying "Meet me in the alley" preceded everything from bicycle races to baseball games to something that ranked as my favorite alley activity of all: fossil hunting.

By the vagaries of geology, the neighborhood where I grew up sat perched on a ridge of sandstone riddled with traces of ancient life. Most of that bedrock lay buried beneath lawns, houses, and forest, but there was an outcrop at a narrow point in the alley, where someone long ago must have blasted to level the roadbed. The result was a steep face of crumbly, beige stone that ran for about thirty feet (ten meters) on the upslope side of the lane—an irresistible target for would-be paleontologists. We quickly learned that chunks of the exposed rock would split apart nicely when hurled onto pavement, sometimes revealing perfect impressions trapped within. Of course, we were hoping to find *Tyrannosaurus rex*, or at least *Triceratops*, and once we even convinced ourselves that a roundish shape in the rock must be the remains of a giant egg. Unbeknownst to us, our dig site dated to a time *after* the fateful asteroid strike that drove non-avian dinosaurs to extinction. But the leaves, twigs, and other plant parts we did uncover told a story far more relevant to the world we were growing up in. They included strange ferns and palm fronds totally unlike the firs, spruces, and pines that now dominate the coastal flora of the Pacific Northwest. Even to the untrained eye, it was obvious that our neighborhood had once looked very different. If we'd had a professional on hand, they could have told us that we'd stumbled across evidence of the Paleocene-Eocene Thermal Maximum, one of the most commonly studied historical analogs of modern climate change.

"It was the warmest time in the past sixty-five million years." That's how geologist and paleobotanist Renee Love describes the period that has become her research specialty. Now an instructor at the University of Idaho, Love wrote her doctoral dissertation about the fossilized plants beneath my hometown in Washington. At 952 pages, it wasn't exactly intended as a field guide, but for my neighborhood visit I'd brought along a copy on my laptop. I hoped that if I could find a few specimens, Love's exhaustive illustrations and photographs would help me put a name to them, adding some scientific context to the species I'd grown up excavating. That plan, neatly worked out in my office at home, came crashing down as soon as I saw the old fossil bed. In place of eroding sandstone, a gleaming white wall of landscaping bricks now ascended the hillside in stark tiers. No trace of bedrock remained exposed. I felt a momentary pang of loss—not just for myself, but also for the current crop of neighborhood kids who would never know the pleasures of a good back-alley fossil hunt. As a practical matter, however, it may have been for the best. When people catch children smashing rocks on the pavement they tell them to go home. The reaction, I had begun to realize, might be rather different for me: a middle-aged stranger in a pandemic mask, carrying a backpack and a hammer. My presence had already drawn a few curious looks; it was time to leave the alley.

Fortunately, I knew of an isolated hillside in the forest south of town where prospecting for fossils wouldn't look quite so conspicuous. It was strewn with chunks of the same kind of rock—Eocene sandstone—knocked loose by erosion from an overhanging cliff. I headed there immediately and soon found two nearly complete examples of something that Love's dissertation told me was an ancient cousin to birch trees.

"That's right," she confirmed, when I reached her later on a Zoom call. "Birches and alders were just evolving from a common ancestor around then. There were quite a few species." As our conversation continued, however, it became clear that birch trees might have been the only familiar sights in that ancient world. Instead of hills and mountains, the local landscape had been spread out flat in a vast network of floodplains, braided stream channels, and oxbow lakes, where a low-lying river system broadened to meet the sea. The climate was wet and subtropical, similar to parts of present-day Mexico. There were tapirs wandering the riverbanks, and huge *Diatryma*—flightless birds that looked like shaggy, thick-necked ostriches with sledgehammer beaks. Crocodiles lurked in the shallows, printing the mud with five-toed tracks and long drag marks from their heavy tails. Renee Love had seen all of these fossils and more, but to her, the exotic animals were a sideshow. The real story—and the link to climate change—lay in the leaves.

"Plants were everywhere," she said, with the easy assurance of someone who has spent a lot of time thinking about fifty-five-million-year-old vegetation. As a graduate student, Love had camped in her van at field sites for months on end, collecting, photographing, and sketching thousands of Eocene fossils. Add to that the long years of analysis, and it's no wonder she could readily describe what had once been a lush forest of giant tree ferns, palms, and, by her count, at least 142 different varieties of leafy trees, shrubs, and vines. Not every specimen could be identified, and some were probably new to science, but she didn't need to put names on the leaves to document their most important features.

"It's called leaf margin analysis," she told me, and described a simple relationship that makes leaf fossils, in her words, "the best way to study past climates." The concept was nothing new. A pair of botanists at Harvard University noted in 1916 that leaves in warm climates tended to have smooth edges, while those in cooler places had edges with lobes or teeth. No one is entirely certain why, though it probably has to do with how plants regulate water loss. (Teeth increase the surface area of leaves, leading to more evaporation through pores at their tips. Fewer teeth may help plants conserve water in warmer climates.) The pattern is

FIGURE 13.1. These leaf fossils date back over fifty million years to the Paleocene-Eocene Thermal Maximum, a time when increased carbon dioxide levels in the atmosphere drove a spike in global temperatures, rearranging ecosystems everywhere. Photo © Thor Hanson.

remarkably consistent across all modern floras, and its discoverers immediately noted that the same principles should apply to the past. "It promises," they wrote, "to afford a simple and rapid means of gauging the general climatic conditions of the Cretaceous and Tertiary." In the century since that first paper, leaf margin analysis has been developed and fine-tuned into a surprisingly precise thermometer for deep time. Given enough fossils, the ratio of smooth to rough leaf edges can pinpoint ancient temperatures to within a few degrees. Other details add nuance, like the prevalence of tapering "drip tip" leaf points so common where rain is warm and abundant. In Love's words, "The paleorecord tells us that climate has changed. Leaf shape tells us how."

With so many leafy fossils at her fingertips, Renee Love had no trouble calculating that the climate of my hometown during the Paleocene-Eocene Thermal Maximum averaged fifteen to twenty-two degrees Fahrenheit (eight to twelve degrees Celsius) hotter than it is today. This comes as no surprise. The whole planet was warmer, a steamy place where temperatures displayed, as Love put it, "very little gradient from the equator to poles." That consistent warmth didn't just alter the vegetation in temperate places like the Pacific Northwest. It allowed subtropical forests to spread as far north as Greenland, and south across the continent of Antarctica. In a hothouse world, there were no glaciers or icecaps to stand in the way.

For climate scientists, the early Eocene provides a telling case study in global warming—not simply because the planet heated up, but because that trend was fueled by greenhouse gasses. Atmospheric carbon dioxide spiked to levels three or even four times as high as those found today, driven by some natural event that experts have yet to agree upon—volcanic activity, perhaps, or a mass release of methane built up in ocean sediments. (A potent greenhouse gas in its own right, methane [CH_4] also adds carbon dioxide to the atmosphere, because as it breaks down, the carbon [C] it releases bonds with oxygen [O_2] already available in the air.) Under a high-emission "business as usual" model, modern climate warming could start reaching Eocene levels by the middle of the next century, making studies like Love's a window into the future as well as the past. When I asked her about the biological implications of repeating that kind of warming, she immediately brought up the concept of mass extinction.

"There have been five major extinction events, six if you count what's happening now," she stated, "and at least half of them can be attributed to climate." Some people might put that figure higher. The first four mass extinctions all occurred at or near various hot and low extremes in global temperature, and even the asteroid incident that wiped out so many dinosaurs probably had less to do with the impact itself than with the sun-blocking dust it kicked up, and the long global winter that followed. Major climate shifts set the stage for extinction by challenging the adaptive capacity of many species all at once, but Love emphasized that mass die-offs weren't inevitable. "Organisms do react," she said, and then rattled off a list of observations from the fossil record: ancient biological responses that could easily have been drawn from earlier chapters of this book. "Life moves around," she said. "Opportunists do well and spread. It's the specialists that tend to suffer." Love had seen all of those things in her data, many times, because the fossils she collected didn't stop at the end of the Paleocene-Eocene Thermal Maximum. They extended into younger Eocene rocks too, when a series of lesser warming and cooling episodes took place. And each time the climate changed, the plant community responded accordingly, shifting its composition of species, with their telltale leaf shapes, but rarely losing entire groups forever. The same can be said for most other early Eocene fossil communities studied to date. Aside from the disappearance of certain bottom-dwelling marine plankton, the take-home message appears to be one of resilience. "The climate changed over and over again," she told me, but the plants and other communities always adjusted. At least, in the long run.

Studying fossils has the advantage of what amounts to a version of time travel. For Renee Love, shifting upward a foot or two in a column of rock could fast-forward her study plants through thousands or even millions of years of history. Compressing time at that scale is a handy approach for answering long-term questions about adaptation and survival—the plants either persist or they wink out. Viewing history from such a distance also offers a sense of detachment from the upheavals in question. This effect can be quite refreshing to those of us caught up in the immediacy of the current moment. "Climate change?" a paleoentomologist once said to me. "It happens every week!" He was exaggerating, of course, but not by much. Scores of examples from the past remind us that global temperatures have always varied, and that plants and animals have responded to every peak and trough, sometimes with resilience, and sometimes with extinctions. But there are limits to what can be learned from fossils alone. Extinction is far from the only measure of ecological upheaval, and even the best deposits are full of gaps (in time and diversity) that blur the details, particularly when it comes to the timing and speed of the events in question.

When inches in bedrock equal eons of time, it's virtually impossible to pinpoint a particular millennium, let alone the minutiae of centuries or decades. Estimates of when the Paleocene-Eocene Thermal Maximum took place, for example, span a range of more than a million years, and while experts agree that the planet warmed rapidly, it's unclear whether that means a span of fifty thousand years or a few hundred. Geologically, the distinction is immaterial, but it makes a tremendous difference in terms of how long plants and animals had to adjust. Finding the best parallels for the rapid change taking place today requires precisely dated fossils matched with an equally accurate climate record. Such combinations may never exist for older events like those from the Eocene—too much time has passed. But dating techniques are better for younger specimens, and when glacial geologists began drilling into the ice sheets of Antarctica and Greenland, they found a way to provide eight hundred thousand years of climate data, not by estimating ancient conditions, but by measuring them directly.

The next time you pour yourself a beverage over ice, take a moment to examine those cubes up close. The one I studied while writing this paragraph was crowded with bubbles that shone in the light of my microscope like tiny silver balls. They weren't there when I filled the trays. We filter our tap water and it looks perfectly clear at room temperature. But all liquid water absorbs gasses from the surrounding air, and when that water freezes, those tiny bits of atmosphere get forced out of the solution and become trapped as bubbles. The air in my ice cube is only a few days old, a snapshot of recent atmospheric conditions in our kitchen. But when bubbles come from deep within a glacier, they date to the exact moment that water transitioned from liquid to crystal. They are, quite literally, tiny samples of an ancient sky, preserving its full range of soluble gasses, including carbon dioxide.

For climatologists, ice cores provide a long-sought record of the link between greenhouse gasses and the global climate, because they also contain a history of temperature. Subtle differences in water molecules reflect the average warmth of the environment where they formed, and that information gets preserved in glacial ice, right alongside those gas-filled bubbles. When data from the two sources are plotted together, they show temperature and carbon dioxide rising and falling through the ages in perfect tandem, like cardiographs of the same heartbeat. (It's one of the ways scientists know that modern temperatures will continue going up, chasing the new peak we've created in atmospheric carbon.) Paleontologists use the ice core data like an almanac, a place where past conditions and trends can be looked up as if flipping through the pages of a calendar. Visible annual layers in the Greenland ice have been counted back sixty thousand years, with chemical analyses taking that number twice as far—it's just the sort of fine-grained precision that researchers needed when searching for past

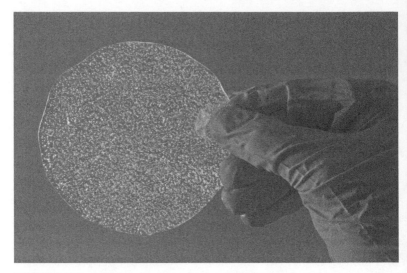

FIGURE 13.2. Bubbles of ancient air are clearly visible in this cross section of an Antarctic ice core. By drilling continuous cores through ice sheets that are nearly two miles (three kilometers) thick, scientists have assembled a climate record dating back eight hundred thousand years. Photo © Pete Bucktrout, British Antarctic Survey.

examples of rapid climate change. Surprisingly, they weren't that hard to find.

At least twenty-five times during the past 120,000 years, temperatures over Greenland spiked by nine to twenty-seven degrees Fahrenheit (five to fifteen degrees Celsius) in just a few decades, and then stayed warm for centuries before gradually cooling off again. Granted, these events weren't driven by carbon dioxide emissions, and they started from a much cooler base temperature, so they don't exactly resemble today's changes. Nor were they truly global in nature—most of them probably stemmed from sudden shifts in currents that funnel warm tropical waters northward in the Atlantic Ocean, limiting their main impacts to the northern hemisphere. But within those affected landscapes and seascapes, plants and animals repeatedly experienced temperature increases at a pace and scale that either met or exceeded what is happening now. Paleontologists and biologists are just beginning to focus attention on these episodes—the high-resolution ice core data are still relatively new. But early results echo a familiar theme: resilience. In a summary of more than sixty recent peer-reviewed publications, researchers found examples of range shifts, behavioral adaptations, and widespread turnover of biological communities, but for nearly the entire time period in question there were few extinctions. What Renee Love had observed at a broad time scale held true in tighter focus: again and again, species and communities showed the flexibility to withstand rapid change—right up until the moment when they suddenly couldn't.

At the close of the last glacial period, as the ice retreated and the planet warmed to roughly its current state, more than 150 species of large animals abruptly disappeared. These extinctions occurred mostly in North America, South America, and Eurasia, taking such famed creatures as the mastodon, the cave bear, and the woolly rhinoceros, as well as lesser-known beasts like that old Joshua tree seed disperser, the Shasta giant ground sloth. Since all of those species had survived similar or even larger temperature swings in the past, experts believe that climate change alone did not trigger the die-off. Aggressive human hunting, the Pleistocene Overkill, is considered the other key factor. Its relative importance remains hotly debated, and probably varied by species and circumstance, but the real lesson lies in the fact of the interaction itself: during episodes of rapid climate change, biological impacts are amplified by other environmental stressors. That insight may help explain why some past climate shifts led to only modest reorganizations of species, while others caused mass extinctions, and why certain habitats or groups were often disproportionately affected (e.g., marine plankton during the early Eocene). When it comes to resilience, context matters—a worrying thought in the current crisis, where natural systems are already coping with human stressors that are far more dire than a few bands of hunters with spears.

As discoveries continue to emerge from the paleorecord, scientists aren't just using them to reassess the past. Many of those lessons apply directly to how we understand, manage, and predict biological climate responses now. For a summary of recent developments, I reached out to Damien Fordham, head of the Global Change, Ecology, and Conservation Lab at the University of Adelaide in Australia. His team focuses on integrating knowledge from ancient systems into modern research and conservation, or, as he has described it, "the intersection of paleoecology, paleoclimatology, paleogenomics, macroecology, and conservation biology." It's a lot to fit on a business card, but Fordham's prolific output suggests that he's found a fertile niche. We exchanged several emails, and he sent me a brand-new paper brimming with ideas that I hadn't encountered anywhere else. Modern conservation strategies, for example, typically ignore fringe populations in favor of protecting habitat at the core of a species' range. But the paleorecord shows that peripheral areas can be critical when the climate changes, acting as focal points for range expansion, and often harboring individuals already accustomed to conditions at the edge of a species' comfort zone. Other innovations include the recovery and analysis of increasingly ancient DNA, and how comparisons with modern samples can identify exactly how climate-driven traits are evolving. Searching fossil collections for viable seeds, spores, or other dormant life-forms is also under way. Sprouted and grown to maturity, they offer the possibility of actually experimenting with older iterations of living species. "The paleorecord provides great opportunities to test and improve warning systems for species extinctions," Fordham summarized, adding that his team is also studying the vulnerability of whole ecosystems. He told me he found "a level of hope" from studying past examples, but he repeated the caution that other human-driven changes have probably eroded nature's resilience.

Buried in a paper that Fordham wrote with several colleagues, I found an offhand comment linked to what may be the most important historical climate lesson of all. They noted that the list of species on the move at the end of the last ice age included, in the parlance of science, "anatomically modern humans." In other words, the bands of hunter-gatherers who followed retreating ice northward in Europe and Asia, and then eventually crossed over to the Americas, did so as part of a climate-driven range shift. They were reacting to their surroundings, taking advantage of new opportunities and looking for their comfort zone in a rapidly warming world, just like any other species. Although we tend to regard human history as something apart from nature, probing our own responses to past climate upheavals is an essential part of understanding and surviving the present one.

For much of the twentieth century, academics steered clear of anything that hinted at what is known as environmental determinism, the debunked theory that certain climates and geographies produce cultures of superior temperament and morality. Colonial powers relied on this intellectually odious notion to justify any number of racist policies, and it has left a lingering stain on the study of relationships between people and their environment. Paleontologists and archaeologists helped rekindle a more neutral interest in people and nature with observations that, like Fordham's, pointed to connections in the distant past. Early hominins first departed Africa for Europe during a cold, arid period, for example, and the development of agriculture in Mesopotamia occurred just as the Near East was getting warmer and wetter at the end of the last ice age. Climate correlations are now being found for well-known events throughout history, from the demise of the Roman Republic (volcanic dust/global cooling) to the rise of Genghis Khan (warmth/moisture/abundant grass) to the French Revolution (drought/crop failure). Researchers carefully avoid giving climatic trends exclusive credit for any particular historical event. It's far more likely that changing conditions worked in tandem with other stressors, just as they did during the rapid extinctions of Pleistocene megafauna. This is not to say, however, that major climate events do not have major consequences for people, and perhaps no period in recorded history makes that point more strongly than the Little Ice Age, a four-hundred-year-long global cold snap that peaked in the seventeenth century.

For climatologists, the Little Ice Age was a relatively minor dip in average temperatures that was triggered and prolonged by an as-yet-undetermined combination of factors—volcanic dust, perhaps, or cyclic changes in ocean currents and solar activity. One controversial theory implicates the colonization of the Americas, when European diseases decimated indigenous populations and led (briefly at least) to the mass reforestation of abandoned farmland. Theoretically, that surge of tree growth could have soaked up a measurable share of the atmosphere's carbon dioxide. Whatever the cause, the long chill left an indelible imprint on human activities around the world, and it did so at a time when—unlike most previous episodes—people were taking notes. Ship's logs, crop reports, explorer's journals, trade accounts, government records, newspapers, diaries, correspondence, and other firsthand accounts of the day were filled with commentary about extreme weather. Like ice cores and fossils to paleontologists, those written records have been a boon to students of human history.

At least four noted scholars have written book-length works about the Little Ice Age in recent years, none more comprehensive than British historian Geoffrey Parker's masterful *Global*

FIGURE 13.3. Ice fairs on the frozen River Thames in London became a regular occurrence during the Little Ice Age, like the one depicted here from the winter of 1683–1684. The festival lasted for several months and was attended by throngs of people who enjoyed such activities as bull-baiting, fox hunts, stilt-walking contests, a horse-drawn ice boat, bowling, pub tents, and more. Image © Museum of London.

Crisis. As the title implies, the book describes a deeply unsettled time, when climate-driven food shortages, floods, storms, droughts, fires, and other calamities helped spur unprecedented levels of conflict. During the 1600s, European powers fended off more than thirty rural uprisings and other revolts, all the while fighting dozens of larger struggles with revealing names like the Nine Years War, the Thirty Years' War, and the first, second, and third English Civil Wars. Collectively, the continent saw only three years of peace in the entire century. In India, the Mughal Empire suffered bloody struggles over succession and fought continuous external wars, on multiple fronts, from 1615 through at least 1707. China endured internal revolts, border skirmishes with Russia and Korea, and a civil war between the Qing and the Ming factions that raged for over sixty years. Beyond these overt signs of strife—what biologists might call heightened aggression—many familiar climate change responses played out. Migration surged, as millions of people moved from farms to cities and then abroad in search of better prospects. Behaviors changed too, from shifts in diet, trade, and the timing and products of agriculture to stranger developments like an uptick in the number of witch trials. (Superstitions of all kind flourished as people sought scapegoats for all the harsh weather.) Again, as in other disciplines, Parker does not paint climate change as the direct cause of these trends so much as a pervasive contributor, an aggravator of existing risks and problems. Military planners use the term *threat multiplier*, which may be the best climate change description I've encountered anywhere—historically, biologically, or otherwise. Scientists have begun picking it up too, and it's also heard with increasing frequency in strategic and diplomatic circles, because, as Parker emphasizes in an epilogue to his book, patterns from the climate-stressed seventeenth century are beginning to repeat themselves in the climate-stressed twenty-first.

From extreme weather to extreme politics, it's easy to find a climate signal running under the surface of recent events, including the 40 percent rise in armed conflicts since the year 2000. The Syrian Civil War, for example, began during the worst drought in that country's history, spurred in part by the desperation of more than a million displaced people who migrated from failing farms to crowded urban centers. And while many pressures led to the Arab Spring uprisings, the critical early protests were sparked by bread shortages, traceable in turn to heat waves and failed wheat crops in Russia and Canada the previous summer. Migration is also on the rise globally, with telling differences between the places that people are leaving and the places they hope to reach.

Studies of both internal and international patterns show measurable migration trends toward relative climate stability and away from landscapes prone to excessive heat, drought, flooding, rising seas, storms, and wildfires.

In its role as a threat multiplier, climate change has become a constant presence in the daily news cycle. While writing this paragraph, I scanned the headlines and quickly spotted stories on record-setting forest fires across the American West, with people stuck inside on "smoke lockdowns" to avoid inhaling the dangerous haze. There was a piece on evacuations ordered in the path of an approaching Atlantic hurricane, and another describing "managed retreat" as a strategy for coastal communities fighting sea level rise. More subtly, a story detailed rising government subsidies for crop failure insurance in India. More overtly, there had been a shortage of air conditioners in Arizona. Seen through the lens of climate change biology, human activities often echo the responses of plants and animals in the wild—moving, adapting, taking refuge. Such parallels are not surprising, because in spite of the complexity of our societies, and the technologies we surround ourselves with, in the end we're just one more species in a changing world, facing the same climate challenges, and drawing on the same basic toolbox of potential solutions. With one notable difference. Unlike any other organism on the planet, people have the ability to do more than simply react to climate change. If we so choose, we can alter the behaviors that are causing it to happen.

Everything You Can

Strong reasons make strong actions.

—William Shakespeare
King John (c. 1596)

B ent low under the open belly of the tractor, I held the newly rebuilt part in position while Noah gave it a few taps with a rubber mallet. It resisted at first, then slid into place with a satisfying thunk.

"We did it!" he said softly, his voice strained with muted excitement. "We conquered the oil pump!"

Talking in low tones had become habitual when we worked in the barn, so as not to disturb the large wasp nest dangling from a nearby rafter. But later, outside in the sunshine, we whooped and gave each other high fives after hand-cranking the engine to life, and watching the oil pressure gauge finally surge up into the safe zone and stay there. The problem had plagued us for months, but now we were ready to push Noah's tractor to the limit. Moments later, sailing down our country lane, we dared fourth gear and hit ten miles per hour, a speed that feels pretty fast on a piece of farm equipment built in 1945.

At first I had assumed that my son's love of antique tractors was a passing phase—lots of young kids like big equipment. But then he started saving up the money he made selling eggs and chickens at our local county fair, and before I knew it I was borrowing a flatbed trailer and transporting an old red Farmall A from someone else's yard to our own. It took us half a year just to get the thing started, a process no doubt hampered by my sense of reluctant irony at simultaneously writing a book about climate change while resurrecting a vehicle with a fuel efficiency rating of four miles to the gallon. But gradually, as Noah and I worked our way through the various repairs, from the magneto to the oil-bath air filter, petcocks, tappets, and more, I developed a grudging admiration for the sheer ingenuity of all those interworking parts. (We also rebuilt the carburetor—several times—but I reserve judgment on the ingenuity of that particular device.). Internal combustion engines bear a lot of responsibility for the climate change crisis, but they are undeniably clever machines, a fascinating way to transform fossil energy into raw power. It came as something of a surprise, then, when I learned that Noah had absolutely no intention of putting that power to work.

Early on in our tractor repair efforts, I spent a bit of time looking around online for implements that would fit the spacing on the hitch and match up with the shaft of the power takeoff unit. I thought Noah might be interested in a good sickle mower and a hay rake—maybe he wanted to add a field-mowing service to his line of agricultural endeavors. But when I mentioned the idea to him, he looked a little startled. Then he patiently explained to me, in the manner of stating the obvious, that his tractor was old and obsolete, and that he'd purchased it as a collectible. The point was not to use it but to fix it up, to make it look and run just like it did when it rolled off the factory floor. He wanted to display it at the county fair and drive it through town in our island's annual Independence Day parade. Then he revealed an even grander plan: taking the tractor to a vintage farming festival, where hundreds or even thousands of like-minded enthusiasts gathered to show off the fruits of their own restoration labors. That's when I realized that refurbishing old, gas-guzzling machines wasn't necessarily a step backward on the path to a low-carbon future. In fact, it may count as a measure of progress. The world will be better off when more of us view the entire era of internal combustion in the same way that Noah and his fellow collectors view their tractors: as history.

For me, that realization provided a helpful nudge, pushing my climate change thinking and fretting finally into the realm

FIGURE C.1. A 1945 Farmall A tractor, up and running. Photo © Noah Hanson.

of personal action. It was on my mind during a visit to our local equipment dealer, as I filled out the paperwork on a new purchase, and then went around to the back to pick it up.

"Looks like it oughta be blowing bubbles," the yard worker grumbled, and I had to admit that he had a point. In the place where an engine would typically sit, the mower featured a dome of white and orange plastic, with a gaping slot for the battery pack. The whole thing felt flimsy and strangely lightweight when we hoisted it into the car. While I liked the idea of cutting grass without burning fossil fuels, this contraption hardly seemed up to the task—particularly for our lawn, which might be more accurately described as a half-tame pasture.

"Are you selling a lot of these?" I asked, and he nodded ruefully.

"Yep," he said. "That's where it's all headed."

Later, his gloomy outlook made perfect sense to me when I saw how well the electric mower worked. It quietly and efficiently trimmed our haphazard yard as well as any gas-powered machine I'd ever owned—and has continued to do so—with no need for oil changes, air filters, spark plugs, new carburetors, or all the other things that make the *repair* side of an equipment business viable. The same can be said for our new chainsaw, our plug-in car, and a host of other battery-powered items my family has switched to in recent years. To be honest, I had been putting off making the transition away from gas and diesel, fully expecting the electric options to underperform. After all, it wasn't too long ago that electric yard tools were notorious for slicing through their own extension cords, and I'd known someone who had an early electric car that required constant recharging—even while driving—with a gas-powered portable generator, which rather defeated the whole purpose. But to my surprise, every piece of modern electric gear we've tried has been a substantial step up from its polluting predecessor, making at least this one small act for the planet a no-brainer.

To be clear, buying electric lawnmowers will not be enough to stop climate change. Even if everyone converted to battery power for their yard work and daily driving, fossil fuels would still be deeply embedded in the global economy, from agriculture and air travel to shipping, construction, and manufacturing (including the production of electric mowers and cars). Nor does a tweak in power use by those people lucky enough to have backyards and vehicles address the complex social and political implications of the crisis, with its gross inequities in cause and consequence. But when faced with a challenge that feels overwhelming, there is power in practicality. I subscribe to a philosophy expressed to me by Gordon Orians, an eminent American biologist whose seven-decade career has spanned everything from blackbird behavior to the evolution of fear. When asked what a concerned citizen should do to combat climate change, his response was immediate and concise: "Everything you can."

In that simple phrase, Orians managed to capture both urgency and agency—the seriousness of the issue, combined with the importance of taking action at a relevant scale. It's not a new idea. Nineteenth-century thinker Edward Everett Hale expressed something similar in a verse conceived long before anyone was worried about climate change. "I cannot do everything, but still I can do something," he wrote. "And because I cannot do everything, I will not refuse to do the something that I can do." The value of the advice from both Orians and Hale lies in their choice of the word *can*, a verb rooted in possibility, and adaptable to any circumstance. It helps us focus our energy on tasks immediately at hand: tangible things like how we drive, shop, eat, travel, protest, vote, and yes, cut the grass. Naysayers will claim that taking personal action is trivial, an empty gesture in the face of a problem so large. But that position is wrong, and not just slightly wrong; it is the opposite of the truth. In nature, we have seen how the responses of individual organisms determine the fates of populations, species, and entire ecological communities. The same pattern applies to society. Addressing climate change requires a fundamental cultural shift in our relationship with energy, from how we produce it to how much of it our lifestyles demand. That makes individual action *more* important, not less so, because it is the collective behaviors and attitudes of individuals that define, and change, a culture. Yes, we need stronger climate policies, and strong leadership to carry them forward, but those things will be the results of cultural change, not the cause of it.

Doing everything one can about climate change is also a fitting approach biologically, because, as this exploration has shown again and again, that is precisely how plants and animals are responding. When faced with a climate challenge, species don't simply give up—they do all that they can to adjust. Some succeed and some fail, and taking the time to learn why gives us new insights into our own reactions. The proliferation of range shifts in nature, for example, tells us something about the upswing in human migration. And the adaptations we observe in fish, bears, and other species alert us to trends in our own behaviors, and how our remarkable plasticity will become more and more important as the planet warms. Models and predictions certainly point to an unsettled, even chaotic future, but nature is filled with examples of resilience that should help inspire us. If butterflies can evolve larger wing muscles in response to this crisis, then shouldn't we at least be able to change a few behaviors—how we drive, for example, or where we set the thermostat? And if lizards can alter the gripping power of their toe pads in a single generation, then perhaps we can find the motivation to skip an unnecessary plane flight, or remember to turn off the lights when we leave a room. Biological responses to climate change are playing out all around us every day. They are a constant, thrumming call to action, and a reminder that we humans are governed by the very same forces affecting plants and animals. What we choose to do now will not just determine what comes next in nature; it will determine our place within it.

For mathematicians, reaching the end of a lengthy proof involves the satisfaction of writing the letters QED, short for the Latin phrase "quod erat demonstrandum," which roughly translates to "it has been shown." Sometimes, I envy that tradition. It evokes a sense of finality seldom experienced in biology, where answering any one question invariably raises several others, an endless cycle that prompts a different nod to the Romans: ad infinitum. That is certainly the case with climate change biology, where the science is developing and expanding right alongside the problem itself—just as globally, and just as rapidly. It's now a running joke that every biologist in the world is studying the effects of climate change; some of them just don't know it yet. Nobody expects anything like a QED moment soon, because the amount of carbon already released into the atmosphere ensures that temperatures will keep rising for decades. (Climate change will be a dominant theme for future biologists too.) Even under a best-case emissions scenario, the impacts of global warming will need to be managed for a long time, and the lives of plants and animals will always be vital guideposts in that effort. Understanding their challenges and responses may not make us worry less about the crisis, but it does help us to worry smart. And that's not a bad starting point—for scientists allocating scarce research funds, for policymakers enacting climate and conservation strategies, and for all of us striving to navigate the ethical and emotional demands of finding a better path forward. It will be a fraught and fascinating journey—for us, and for every species. I hope we get it right.

Acknowledgments

B eneath its veneer of solitary toil, the creation of a book is very much a group effort. Conceiving, researching, writing, and producing this volume involved contributions from people all over the world, connected only by their willingness to help nudge a lengthy project along its path. I'm indebted, as always, to my friend and agent Laura Blake Peterson at Curtis Brown, and I'm grateful that we once again found ourselves working with the inimitable Thomas Kelleher at Basic Books. The whole team at Basic has been terrific, including Lara Heimert, Rachel Field, Laura Piasio, Melissa Veronesi, Liz Wetzel, Kait Howard, Jessica Breen, Kara Ojebuoboh, Melissa Raymond, Abigail Mohr, Caitlyn Budnick, Mike van Mantgem, and no doubt many others behind the scenes. My appreciation abounds for all the booksellers and librarians out there helping ideas find an audience, with a special nod to Heidi Lewis for her masterful use of the interlibrary loan. Finally, I live in daily thanks for the love and support of my wife and son, and for the many other family members and friends willing to put up with my muttering eccentricities.

Here, in no particular order, is a list of other generous folk who have contributed their time, knowledge, and enthusiasm to this project—I thank them all (and apologize to anyone inadvertently omitted!). Sophie Rouys, Drew Harvell, Nina Sottrell, Robert Michael Pyle, Nicole Angeli, Ann Potter, Richard Primack, Steve and Donna Dyer, Phil Green, Peter Dunwiddie, Barry Sinervo, Dan Roby, Staffan Lindgren, Ben Freeman, Bill Newmark, Victoria Peck, Will Beharrell, Thomas Alerstam, Gretta Pecl, Will Deacy, Songlin Fei, John Turnbull, Sandy Reid, Melissa McCarthy, Sally Keith, Elias Levy, Colin Donihue, Alicia Daniel, Elizabeth Thompson, Constance Millar, Libby Davidson, Bryant Olsen, Carla Laurenço, Simon Evans, Lars Gustafsson, Cody Dey, Anton Mostovenko, Chad Wilsey, Chris Shields, W. Robert Nettles, David Grémillet, Ken Cole, Randy Kolka, Layne Kennedy, Amanda Kueper, Keith Goetzman, Ryan Kovach, Jonathan Armstrong, Renee Love, Gordon Orians, Damien Fordham, Brooke Bateman, and Montague H. C. Neate-Clegg.

Glossary

acclimatization—a form of rapid adaptation to environmental conditions, made by individuals through their inherent physical or behavioral abilities.

adaptation—changes in an organism in response to its environment. Adaptation can be immediate through behavior or a built-in capacity (see plasticity), or evolutionary through the inheritance of adaptive traits.

aragonite—a mineral made up of calcium carbonate; common in the shells of marine life, but less stable than calcite.

backcrossing—breeding between a hybrid and one of its parent species.

biodiversity—the variety of life, including the number of species, the genetic variation within them, and the complexity of the communities they form.

calcite—a white mineral made up of calcium carbonate; common in the shells of marine life.

carbonic acid—a weak acid formed when carbon dioxide dissolves in water.

cascading effects—ecological consequences set off as a chain reaction from a particular event, activity, or change.

chromosome—a structure within cells that carries genetic information.

copepod—one of a large and diverse group of tiny aquatic crustaceans.

coral bleaching—when heat-stressed corals expel their colorful, symbiotic partners, the process leaves them weakened and often pale or "bleached."

Cretaceous—a period of time stretching from 146 to 65 million years ago, characterized by warm climatic conditions, the emergence of flowering plants, and the long dominance of dinosaurs.

critical thermal maximum—the temperature above which an organism ceases to function; lethal temperature.

dinoflagellates—single-celled aquatic organisms related to brown algae; many are capable of photosynthesis, and some live as symbiotic partners within corals.

El Niño—an irregular change in the location and extent of warm surface waters in the eastern Pacific; associated with widespread changes in oceanic and climatic conditions.

Eocene—the second epoch of the Tertiary Period, lasting from roughly fifty-six to thirty-four million years ago.

genetic drift—evolutionary change through random inheritance.

genus—a taxonomic group of closely related species.

heliotherm—an animal that regulates its body temperature by basking in sunlight.

hybridization—crossbreeding between genetically distinct species or subspecies.

hydrology—the study of water and its relationship to the environment.

introgression—the movement of genetic material between species or populations through hybridization and repeated backcrossing.

isotope—one of multiple forms of the same element that are equivalent chemically but differ in atomic mass because of the number of neutrons in their nuclei.

megafauna—large-bodied animals; can be used to describe modern animals (e.g., elephants, bison), but is often used in reference to species that went extinct at the end of the Pleistocene (e.g., mammoths, giant ground sloths, saber-toothed cats).

meiosis—cell division that produces gametes (e.g., sperm and eggs) containing half the genetic material of the parent cell; a precursor to sexual reproduction; also known as reduction division.

methane—a flammable gas with the chemical formula CH_4; the main constituent of natural gas.

midden—a pile of refuse or debris; in biology, often used to describe the nests of hoarding rodents like pack rats.

mutation—a random change in the genetic code of an organism; one of the major sources of variability in nature.

mutualism—a relationship between two species where each partner benefits from the interaction.

natural selection—colloquially known as survival of the fittest; an evolutionary process where the best-adapted individuals survive to pass on more of the genetic material that made them so.

nonlinear—literally "not in a straight line"; applied to relationships in science with inherently unpredictable outcomes.

osmotic shock—the stress of passing suddenly between two fluids of starkly different densities or chemical compositions.

Paleocene-Eocene Thermal Maximum—a warm period approximately fifty-five million years ago characterized by high levels of atmospheric carbon dioxide and a hothouse climate.

pathogen—a virus, bacteria, or other tiny organism that causes disease.

periostracum—a thin, varnish-like organic layer that surrounds and helps protect the shells of various snails, clams, pteropods, and other shell-making organisms.

phenology—the timing of seasonal events in nature.

photosynthesis—the process by which plants and some other organisms use sunlight to transform carbon dioxide and water into carbohydrates.

plasticity—built-in adaptability; an organism's inherent capacity to respond to its environment.

Pleistocene—a geological epoch lasting from 2.6 million years ago to 10,000 years ago, characterized by repeated large-scale glaciations or "ice ages."

polyp—a life-form of certain ocean invertebrates like corals or anemones, typically with a mouth and tentacles set atop a columnar body.

pteropod—any member of the group Pteropoda, free-swimming ocean snails known variously as sea butterflies or sea angels.

quartzite—a rock type largely composed of the mineral quartz; usually formed from sandstones transformed by the geological process of extreme heat and pressure known as metamorphosis.

refugium—a location resistant to environmental change that provides refuge to formerly common species driven from the surrounding area by new, unfavorable conditions.

selection—in evolution, a process that helps determine which traits are passed down from one generation to the next (e.g., natural selection, sexual selection).

sexual selection—in short, selection through mate choice; the idea that preferences and competition in the context of mating determine the inheritance of related traits.

symbiont—a partner in a symbiotic relationship.

symbiotic—a biological interaction where two dissimilar organisms live in close association, usually to the benefit of one or both partners.

talus—boulders and rocks of all sizes that accumulate on the slope below an eroding cliff.

tapir—a plant-eating rainforest mammal that resembles a large pig, but more closely related to a horse.

Tertiary Period—a period of time that lasted from 65 to 2.6 million years ago, characterized by the rise of many familiar groups of plants and animals, including mammals.

timing mismatch—a climate change conundrum where dependent organisms (e.g., plants and their pollinators) alter their schedules differently in response to new conditions, reducing or eliminating the critical time period when they would normally interact.

yucca—a genus of fifty to sixty spiky-leaved desert succulents from North and South America, of which Joshua trees are by far the largest member.

zooplankton—tiny aquatic organisms that drift with the current or have limited swimming abilities, including various protozoans, crustaceans, and the larval forms of many fishes and other marine creatures. (Different from phytoplankton, which are tiny plants.)

zooxanthellae—the various dinoflagellates that live within corals.

zygacine—a toxic alkaloid found in all parts of the death camas plant.

Notes

Introduction: Thinking About It

xi **"I am thinking, brother . . .":** *King Lear*, Act I, Scene 2; Bevington 1980, p. 1178.

xv **chemicals that demonstrably change the way we think, feel, and remember:** Among several neurotransmitters associated with the brain's response to stories, the hormone oxytocin is perhaps the best studied. Its relationship to feelings of empathy and trust has led one researcher to dub it "the moral molecule" (Zak 2012). The oxytocin and other chemicals released when our brains process stories are thought to enhance understanding and help transform abstractions into action.

Part One: The Culprits (Change and Carbon)

1 **"If you want to make enemies, try to change something.":** Wilson 1917, p. 286.

Chapter One: Nothing Stays the Same

3 **"All change in habits of life and of thought is irksome.":** Veblen 1912, p. 199.

7 **"what is . . . is.":** Burnet 1892, p. 185.

7 **". . . fixed limits, beyond which they cannot go,":** Linnaeus made this comment in his 1737 *Critica Botanica* to distinguish

true plant species—those ordained by God—from the many artificial varieties developed by florists. The latter represented the "infinite sport of Nature," but he believed they were always ephemeral, eventually reverting to their true form. Hort 1938, p. 197.

7   **". . . always will be exactly the same.":** ibid.

9   **"succession of worlds,":** Hutton 1788, p. 304.

9   **"may be deemed rare or extinct.":** Jefferson 1803.

10 **"zealous disciple":** Darwin wrote about his love of geology in an 1835 letter from the *Beagle* to his cousin, the clergyman and naturalist William Darwin Fox. In it, he noted his admiration for Lyell's ideas, and said geology offered "so much larger a field for thought" than the other branches of natural science. Though eclipsed by his later work on evolution, Darwin's observations on South American geology and the formation of reefs and atolls were enough to earn him the Wollaston Medal, the highest honor awarded by the Geological Society of London. See Herbert 2005; Darwin Correspondence Project, "Letter no. 282," accessed on September 3, 2018, www.darwinproject.ac.uk /DCP-LETT-282.

11 **"They abound with active Volcanoes.":** Darwin Correspondence Project, "Letter no. 282," accessed on September 3, 2018, www.darwinproject.ac.uk/DCP-LETT-282.

12 **"the most obvious and gravest objection which can be urged against my theory.":** Darwin 2008, p. 279.

12 **"the extreme imperfection of the geological record.":** ibid., p. 279.

12 **". . . and of each page, only here and there a few lines.":** ibid., p. 303.

13 **and supporters began applying it broadly:** Some might say too broadly. Stephen Jay Gould later expressed bemusement at how punctuated equilibrium was being invoked to explain everything from the history of language to the spread of new technologies. While similar patterns of rapid change and stasis may be common, he and Eldredge intended their theory to explain only the life spans of individual species in the context of macroevolution.

14 **"calamities for future generations,":** Von Humboldt and Bonpland 1907, p. 9.

14 **"vast amounts of gas and steam"**: Von Humboldt 1844, p. 214. Translation by Nina Sottrell, personal communication.

14 **"upon which civilization has no significant influence."**: ibid.

15 **"consumption of coal, petroleum, etc."**: Arrhenius 1908, p. 58.

15 **"any doubling of the percentage of carbon dioxide in the air . . ."**: ibid., p. 53.

15 **help stave off the possibility of another ice age**: It was an interest in glacial cycles—a popular topic of scientific debate at the time—that led Arrhenius to derive his famous climate calculations. He focused primarily on how decreased levels of atmospheric carbon dioxide could explain past ice ages and potentially trigger a new one, something he viewed as an existential threat that would "drive us from our temperate countries into the hotter climates of Africa." Arrhenius 1908, p. 61.

15 **on track to double atmospheric carbon dioxide in three thousand years**: Arrhenius had a bit of a showman's flourish, and he imparted this particular detail of his theory not in a staid scientific paper but in a popular lecture given at Stockholm University in January 1896. Crawford 1996, p. 154.

Chapter Two: Mephitic Air

17 **". . . and strive to make measurable all that is not . . ."**: Versions of this quotation are often mistakenly attributed to Galileo directly. It comes instead from an interpretation of Galileo's approach to science, as written by one of his biographers, French scholar Thomas-Henri Martin. Martin 1868, p. 289; translation by S. Rouys, personal communication.

18 **"an intellectual streak of legendary proportions."**: Johnson 2008, p. 41.

18 **". . . in the neighborhood of a public brewery."**: Priestley 1781, p. 25.

19 **". . . any kind of substance may be very conveniently placed."**: ibid., p. 25.

19 **"large, strong frog."**: ibid., p. 36.

19 **"a peculiar taste."**: ibid., p. 35.

20 **"pleasant acidulous taste."**: ibid., p. 28.

20 **This breakthrough earned Priestley the prestigious Copley Medal:** Priestley's discovery of carbonated water was initially met with great excitement from doctors in the Royal Navy, who mistakenly considered it a potential cure for scurvy. Then, as now, medical advancement was seen as a key justification for the funding and pursuit of science. Joseph Black had begun his work on carbon dioxide in search of a remedy for bladder stones.

21 **fermentation occurs in a lot more places than vats of beer:** Though commonly associated with microbes, fermentation is a fundamental and widespread metabolic activity. Human muscles use it when oxygen runs low in the blood, which is why lactic acid often builds up and causes cramping for runners and other athletes toward the end of a long race. Yeasts also rely on fermentation, and the carbon dioxide given off is what makes bread dough rise, forming those little bubbles that will later trap melted butter and jam on a piece of toast.

22 **(Gasses get warmer under pressure.):** The relationship between pressure and heat is familiar to many, ironically, from the workings of the internal combustion engine. Inside each cylinder, a piston compresses air mixed with fuel to a very high temperature. For gasoline engines, the ignition then comes from a spark plug, but in diesel engines, the combustion relies on the heat of compression alone.

Chapter Three: Right Place, Wrong Time

35 **"that early yellow smell":** Wisner 2016, p. 24.

35 **winter conditions matter too:** Primack's team found that Walden winters are also heating up, but in years when January temperatures remained cold, some plants responded with delayed flowering times, no matter how warm the spring. Other studies have found links between fall conditions and spring flowering, suggesting that changing conditions year-round can influence biological events in any particular season. See Miller-Rushing and Primack 2008.

35 **Average temperatures around Walden Pond have risen 4.3 degrees Fahrenheit:** Warming at Walden exceeds the global

average increase of 1.4 degrees Fahrenheit (0.8 degree Celsius) over a similar period, demonstrating that some parts of the planet are warming far more rapidly than others. Walden's rise has also been influenced by urbanization in and around nearby Boston—the loss of vegetation and the spread of heat-absorbing pavement and buildings create a "heat island" effect, making cities measurably warmer than the surrounding countryside.

36 **"the sweltering inhabitants of Charleston . . . drink at my well.":** Thoreau 1966, p. 197.

37 **When the climate was stable, this trait didn't count for much:** It's possible that responding quickly to temperature change was even detrimental before the climate started warming. "New England has the most variable weather of any temperate forest in the world," Primack told me, and explained that early bursts of warm weather could easily be followed by another freeze or snowfall. He speculated that the more conservative plants might have once benefited from being cautious. "They're like New Englanders," he joked. "They don't want to be fooled!"

37 **"seven miles":** Thoreau 1966, p. 103.

38 **"Nature's grandest voice,":** Thoreau 1906, p. 349.

38 **birds continue to show up on the same schedule they used in Thoreau's day:** Comparing Thoreau's bird observations with modern data does reveal one climate-related impact on migration: some birds are no longer bothering with it. Fox sparrows, purple finches, and other species that used to flee at least short distances southward to escape harsh winters can now live comfortably at Walden year-round. Thoreau would also have been surprised to see my red-bellied woodpecker, a southern species whose range has shifted hundreds of miles northward in recent decades. While backyard birdfeeders and other suburban opportunities may have contributed to such trends, studies point to milder temperatures as the dominant factor. See Kirchman and Schneider 2014.

39 **computers at a climate monitoring station automatically deleted the data as false:** See Fritz 2017.

41 **one local bee species figured out a solution:** Several varieties of death camas occur across a wide swathe of the American West,

always accompanied by the death camas bee or, in some locations, a hover fly with a similar knack for detoxifying the alkaloid zygacine.

Chapter Four: The Nth Degree

45 **"Climate is what you expect; weather is what you get.":** This quotation is often erroneously attributed to Mark Twain, or to science fiction writer Robert A. Heinlein. Heinlein included it as part of a list of aphorisms in his 1973 novel, *Time Enough for Love*, but the saying had already been around for decades. It probably traces to a similar observation made by an anonymous schoolchild in the 1887 collection *English as She is Taught: Genuine Answers to Examination Questions in our Public Schools*. That rendition reads, "Climate lasts all the time and weather only a few days." (Le Row 1887, p. 28.) Twain quoted it in his enthusiastic review of the book, which appeared the same year in *The Century Magazine*.

51 **(or at least something virus-sized):** It is extremely difficult and time-consuming to isolate and culture pure viral strains, but Harvell and her colleagues determined that viral-sized samples from sick starfish could infect healthy ones. They also found the DNA of a likely viral suspect in their samples—a close relative of the canine parvovirus deadly to dogs. See Hewson et al. 2014.

52 **making its pathogens more prolific:** Many viruses, bacteria, and other pathogens thrive in warmer environments, a reminder that there are often winners and losers in relationships affected by climate change. While it's a poor time to be a starfish, it's a great time to be a starfish virus.

53 **the concept and phrase "keystone species.":** Paine initially defined a keystone species as a mid- or high-level predator that maintains community structure and diversity by suppressing populations of its prey species. Use of the term has since broadened to include any species that exerts an unduly large influence on its native ecosystem.

Chapter Five: Strange Bedfellows

57 **"Misery acquaints a man with strange bed-fellows. . . .":** *The Tempest*, Act II, Scene 2; Bevington 1980, p. 1511.

60 **where seabirds like tufted puffins formerly nested:** The fleeting overlap between tufted puffins and brown pelicans reflects an increasingly common climate change dichotomy. The same warming that allows pelicans to advance from the south is apparently making things too hot for puffins. Though still common in coastal Alaska, where temperatures remain within their comfort zone, puffins have declined dramatically throughout the southern part of their range. Once common in Washington State, they were added to the local endangered species list in 2015. Tellingly, that was the same year that brown pelicans—once rare—became common enough to be removed from the list.

61 **leaping an average of 215 miles (345 kilometers) north in only four years, including barnacles:** While 215 miles might sound like a tough trudge for a barnacle, they can travel long distances on ocean currents during their larval stage. Many other marine organisms also disperse as larvae, including various mollusks, anemones, crustaceans, bryozoans, tunicates, echinoderms, and fish.

63 **their ability to infest and overcome perfectly healthy trees:** Staffan Lindgren believes that mountain pine beetles evolved this unusual habit to escape competition. By attacking live trees with the help of fungi, they open up a huge food and habitat resource unavailable to other bark beetles. While they can also be found alongside their cousins in dead or dying trees, they aren't particularly successful in that highly competitive setting. In Lindgren's words, "They just sort of hang in there." See Lindgren and Raffa 2013.

65 **the Lindgren funnel trap is a device he invented as a graduate student:** "It was a nightmare," Lindgren said, describing the old method of catching beetles on hardware cloth coated with sticky goop. The stuff got all over clothes and hair and required hours of processing with solvents to free the specimens.

"Being basically lazy, I thought there had to be a better way!" he laughed. That lazy impulse inspired an ingenious treelike arrangement of stacked funnels. Baited with the appropriate pheromones, this trap can be used to draw in a wide variety of the world's six thousand bark beetle species—the curious beetles eventually slip while exploring the funnels, tumbling down to a collecting jar located at the bottom.

66   **"the extreme tameness of the birds":** Darwin 2004, p. 355.

67   **looks more or less like a mouse turd:** Entomologist Diana Six at the University of Montana often begins presentations about the mountain pine beetle outbreak with an image of the beetles and mouse droppings side by side.

67   **"plan for surprise.":** as paraphrased in Cooke and Carroll 2017.

Chapter Six: The Bare Necessities

69   **"Unseen in the background, Fate . . .":** Wodehouse 2011, p. 186.

71   **to examine one small piece of that puzzle:** My research involved hiding hundreds of wicker nests in the forest and baiting them with soft clay eggs that preserved the tooth marks of would-be attackers. It worked well enough to fool rodents, and I found their bucktoothed bites at similar rates in forests of all sizes—matching the ubiquitous distribution of rats and mice in the study area. But Bill and his crew later followed up by locating and studying the fates of over a thousand real nests and found that indeed most birds suffered higher rates of nest loss in forest fragments, presumably from increased attacks by raptors, snakes, and other non-rodent predators. See Newmark and Stanley 2011.

73   **But the Freemans' results suggested the opposite:** The Freemans and others point out that upslope migration is not only prevalent in the tropics, it is more consistent, with whole suites of species showing the same patterns. Things are changing on temperate mountains too, but the process is more idiosyncratic and case specific. Freeman told me part of the reason probably has to do with seasonality. Temperate species have the ability

to react in time as well as space—breeding earlier in the springtime, for example, rather than moving to a new location. The
tropics lack that variability. "If tropical birds want a certain climate, they have to go and find it."

74 **but the mosquitoes that carry it—like birds—are moving
uphill:** Like people, birds suffer from a range of insect-borne
diseases that follow along as their vectors respond to climate
change. Avian malaria is expanding into warming habitats just
like human malaria, and has already been linked to the decline
and upslope migration of several rare honeycreepers in Hawaii.
See Liao et al. 2017.

77 **something mysterious about what specific changes cause a species to move:** Upward migration is sometimes explained as a
competitive chain reaction, with lower-elevation species driving
others ahead of them as they climb. That may happen in some
cases, but the process is usually far more subtle and complex. In
Peru, for example, not a single one of the birds that Ben Freeman
found absent or dwindling from their ridgetop had a direct com-
petitor ascending from below.

78 **Oceans and lakes interact with the atmosphere in the same
way:** In addition to mixing with air at the surface, lakes receive
carbon dioxide from decomposing leaves, wood, and other organic materials that wash in from surrounding landscapes, making it harder to study the effects of climate-related acidification
directly. (See Weiss et al. 2018.) The ocean carbon cycle is also
more complex than atmospheric exchange alone, but it's comparatively easy to distinguish the various sources.

79 **slowly dissolving a duck egg in a glass of seltzer:** Sparkling bev-
erages of all sorts are mildly acidic, but seltzer is a good choice to
test the effects of carbonic acid because it lacks the neutralizing
salts found in club soda, or the competing acids and ingredients
common in sugary drinks like colas. We used a duck egg, but a
chicken egg would work just as well. If you happen to have a tur-
tle egg, that would be ideal, since turtles use aragonite for their
shells, just like baby oysters and sea butterflies. Bird eggshells
are made of sturdy calcite, but since seltzer is far more acidic
than seawater, it still got the job done, leaving us with an egg

enclosed only by its translucent, rubbery membrane. The whole process took seventeen days, and we changed the water periodically to replenish any acidity that may have fizzed away. It's a great demonstration of how carbonic acid erodes shells, and also a good use for that old seltzer in the pantry that you ended up with after a potluck, and that no one in your family will drink.

79 **acidic water turns carbonate into bicarbonate:** More specifically, carbonic acid breaks down into hydrogen ions and bicarbonate. Those hydrogen ions then bind with carbonate already in the seawater to form more bicarbonate. And when supplies of free carbonate run low, the hydrogen ions begin dissolving shells to get at more. The result is an increasingly corrosive environment with little free carbonate available for shell formation or repair.

Chapter Seven: Move

85 **"'I'll move the world,' quoth he.":** Mackay 1859, p. 151.

86 **fanciful notions from antiquity:** Aristotle shared the belief that swallows hibernated in winter, and he thought that songbirds like redstarts and warblers transformed themselves seasonally into other species. One of the more peculiar ancient theories held that cranes migrated to the Upper Nile region, which is partially true, but added the bizarre twist that they spent the winter months in pitched battle with armies of pygmy warriors mounted on goats. This theme was repeated in artwork, stories (e.g., Homer's *Iliad*, *Aesop's Fables*), and scientific treatises (e.g., Aristotle, Pliny, Aelian) for over a thousand years. See Ovadiah and Mucznik 2017.

86 **Swedish taxonomist Carl Linnaeus and one of his students:** Though he prided himself on his knowledge of birds, Linnaeus was best known for his botanical skills. The 1757 treatise on migration, *Migrationes Avium*, is often attributed to Linnaeus alone, but it was actually a summary of the work of one of his students at Uppsala University, Carolus Daniel Ekmarck. It's generally assumed by scholars that Ekmarck and Linnaeus wrote the paper together. See Heller 1983.

86 ". . . at the commencement of the pleasantest part of the year.": Ekmarck 1781, p. 237.

86 "When the frost rages and the tempests beat.": White 1947, p. 60.

87 ". . . from south to north according to the season.": ibid., p. 124.

87 "etc., etc.": ibid., p. 124.

88 "love and hunger,": ibid., p. 129.

88 "enjoy a perpetual summer,": ibid., p. 162.

88 "retreat before the sun as it advances.": ibid., p. 124.

88 he would no doubt marvel at the various new techniques for tracking animals: Gilbert White might also be startled to learn that his quirky little book about Selborne remains in print, one of the better-selling titles in the history of publishing after nearly three hundred different editions.

88 most species crave familiarity: Even scientists have been known to follow their comfort zone to new landscapes. After the United States elected a fact- and science-averse president in 2016 and pulled out of the Paris Climate Agreement, dozens of top American academics and students accepted an invitation from French president Emmanuel Macron, moving to the welcoming atmosphere of France under a new $70 million climate research program dubbed "Make Our Planet Great Again."

92 Beyond these general patterns: A related but less well-known pattern is developing at sea, where cool-water species are descending to greater depths as the surface warms. This affects home ranges but also short-term trends known as vertical migrations. Antelope on the Serengeti get all the press, but by far the largest mass migration on the planet is the daily movement of plankton up and down in the water column.

93 "Birnam Wood to high Dunsinane hill shall come against him,": Macbeth, Act IV, Scene I; Bevington 1980, p. 1239.

93 "leafy screens.": Macbeth, Act V, Scene 5; Bevington 1980, p. 1247.

97 This echoes a study from California: Crimmins et al. 2011.

98 ". . . would take something like a million years.": Reid 1899, p. 25.

98 "... and that are killed by digestion,": Reid 1899, p. 28.

98 **Reid's Paradox.:** Paradoxes were apparently a family affair. Reid's great uncle, physicist Michael Faraday, had his name attached to an even more famous conundrum in the field of electrochemistry.

100 **the comparatively stately pace of just over a half mile (one kilometer) every decade:** This figure is the average for 254 birds tracked from 1974 to 2005. The authors looked at changes in both the leading and trailing edges of range shifts as well as the "center of abundance," a similar metric to Fei's "geographic center." See La Sorte and Thompson 2007.

Chapter Eight: Adapt

101 **"Nature's verdict is . . .":** Carver 1915, p. 74.

103 **it's still a diet way too high in protein:** Bears aren't the only ones who have trouble putting on the pounds (or keeping them on) when they eat too much protein. Weight loss schemes from the Atkins diet to Sugar Busters to the South Beach Diet all rely on a high-protein model. People following the Stillman diet, for example, are getting 68 percent of their calories from protein, more or less like bears at a salmon stream.

108 **unconscious physical adjustments . . . made life in the California sunshine feel perfectly normal:** Another familiar example of human plasticity involves life at high altitudes, where the body compensates for reduced oxygen in the air by producing additional red blood cells and adjusting breathing patterns, heart rate, blood pressure, and more. Athletes are thought to gain temporary advantage by training or sleeping in such conditions because it boosts their oxygen uptake when they return to lower elevations to compete. This is why the main United States Olympic training facility is located in the Rocky Mountain foothills, why many European teams train in the Alps, and why elite athletes in Australia—the world's least mountainous continent—spend time in climate-controlled "altitude houses," with an artificial atmosphere that mimics conditions above 9,800 feet (3,000 meters).

108 **adult body size is based in part on cues received during the first stages of development:** In general, 80 percent of a person's height is controlled by their genetic heritage, with the final 20 percent attributed to plasticity and environmental influences. There is a lot of variation, however, since at least fifty different genes are involved! See McEvoy and Visscher 2009.

108 **sufficient food and other resources necessary to support a larger size:** This intriguing link has been shown in everything from birds to hamsters to *Homo sapiens*. Stress during development and earliest childhood leads to smaller stature, lower metabolism, and a range of other physiological differences. All are thought to be adaptive for harsh conditions where food may be in short supply, and where it would be challenging for an individual to maintain a larger body. Efficient, smaller bodies can become problematic, however, if the environment changes substantially later in life. Smaller individuals can suffer health consequences in processing abundant food, for example, and may struggle to compete against larger individuals more suited to exploiting a plentiful resource. See Bateman et al. 2004 for a fascinating discussion of how this kind of plasticity impacts human disease prevalence, including a putative link to type 2 diabetes.

109 **making the other critically endangered:** To add insult to injury, California dandelion flowers readily accept the pollen of their more common cousins, exposing the species to the additional risk of being genetically swamped into oblivion through hybridization.

110 **after the water warmed substantially in 2009 and 2010:** This marine heat wave was associated with an El Niño event, something that climate scientists expect to increase in frequency and intensity as the planet warms. Interestingly, being exposed to such cycles in the past may help explain why Humboldt squid evolved and maintained such dramatic plasticity, a trait now expected to help them weather the challenges of climate change. See Hoving et al. 2013.

113 **butterfly fish become docile in a bid to save energy:** Keith believes that the fish may be diverting energy from reproduction as well as aggression, which would explain why their numbers

don't drop off until several years after a bleaching event. Docile, energy-saving behavior allows the adults to struggle along, but when they finally start dying of old age there isn't a new generation around to replace them.

114 **plasticity (useful when things change):** In theory, plasticity gives species an edge wherever the environment is variable. There is some evidence for this in terms of latitude, with plants and animals in the temperate zone often (but not always) showing higher plasticity than their relations in the tropics. This pattern presumably evolved because higher latitudes experience larger annual seasonal extremes, and they have a long history of climate upheavals from glacial events during the Pleistocene.

Chapter Nine: Evolve

117 **"If we want things to stay as they are, things will have to change.":** Di Lampedusa 1960, p. 28.

122 **mean spiders fare better than friendly ones:** The species in question lives in colonies on large webs strung over rivers and streams. Experts have yet to determine why aggressive colonies prosper and pass on their traits after storms, but it may have to do with efficiency in subduing limited prey, or the ability to drive off competitors. See Little et al. 2019.

123 **a messy version of this lesson will take place right on your stovetop:** There is another fundamental climate change lesson with a clear analogy in the kitchen. It has to do with sea level rise. While melting ice sheets in Antarctica and Greenland are adding to rising seas around the globe, over half the increase to date can be attributed to heat. Simply put, warmer water takes up more space, so as the oceans rise in temperature, they also swell in volume. Demonstrating this principle (in reverse) involves a culinary experiment that is very straightforward, but nonetheless difficult for some of us to pull off: pour a cup of hot coffee and then forget to drink it. If you can do this, and if you remember to measure the depth of the coffee while it's hot, you'll notice something different when you return to that cold, forgotten cup. In my trial, the coffee level dropped over a

quarter of an inch (seven millimeters) as it cooled, enough to make it look as if someone had taken a considerable sip. While some of that lost volume disappeared as steam, much of the difference was caused by the drop in temperature alone. (If you plan to go ahead and drink the cold coffee, as I did, then I suggest using a metal ruler. It turns out that wooden rulers are covered with varnish that melts off in hot liquids, which I only realized later must have been the source of the coffee's unpleasant tanginess.)

125 **In short, the females preferred them:** There is a long-standing debate among evolutionary biologists about the importance of sexual selection, and what drives it. Is the process simply a matter of preference, beauty for the sake of beauty? Or are the preferred traits linked to some underlying measure of health, essentially an advertisement for parental fitness? There is evidence to support both views, and they needn't be mutually exclusive, but it's probably safe to say that such uncertainties have often kept sexual selection out of the evolutionary limelight. For a vigorous exploration of the topic, see Prum 2017.

126 **algae-rich waters have obscured visibility at many coastal mating grounds:** Like so many other climate change conundrums, the challenge for mating sticklebacks is exacerbated by other human-induced problems. In addition to warmer water, algae growth is fueled by nutrient-rich runoff from agriculture, sewage, and other land-based activities.

127 **drift becomes far more powerful when populations shrink and become isolated:** The effects of genetic drift on small populations are generally thought to be negative, because when randomness overpowers the forces of selection, it can allow harmful mutations to accumulate. In other words, traits that would normally be selected against and removed from the gene pool are more likely to endure. Yet in nature, many species persist in small populations indefinitely. A new theory addresses this inconsistency. See LaBar and Adami 2017 for a fascinating discussion of what they call drift robustness, how negative and positive mutation rates can sometimes reach an enduring mathematical balance.

129 **". . . but cannot stop himself from doing so.":** Montana
Conservation Genetics Lab website: www.cfc.umt.edu/research
/whiteley/mcgl/default.php.

130 **In plants, the results often create novel evolutionary pedigrees:**
There are many reasons why plant hybrids tend to be more enduring than animal hybrids. One of the most important has to do
with chromosome number. When parental species possess different numbers of chromosomes, their hybrids often inherit an
odd number that renders them infertile. Horses have sixty-four
chromosomes and donkeys have sixty-two, for example, so the
poor mule ends up with sixty-three. That prevents its eggs and
sperm from developing properly, because sixty-three can't be divided equally during meiosis, the form of cell division necessary
for sexual reproduction. The same thing happens in plants, but
there's a twist. Many plants can reproduce vegetatively, allowing
infertile hybrids to persist, and their chromosomes are also prone
to spontaneous doubling, making an odd-numbered hybrid suddenly viable. For a fascinating review of this topic, see Hegarty
and Hiscock 2005.

132 **Plant hybrids are often fitter than their parent species:** A phenomenon known as hybrid vigor often makes first-generation
crosses particularly robust. It's not entirely understood why
this occurs, but it's generally thought to stem from increased
heterozygosity—the way disparate parents contribute broader
genetic variation for any particular trait. The effect usually fades
quickly in subsequent generations, however, as descendants of
the original cross reproduce amongst themselves and their genes
become more homogenized. Gardeners and farmers are familiar
with a version of this concept via their annual purchase of hybrid seed varieties, which tend to produce large and productive
plants initially, but don't reliably pass on those characteristics
to their offspring (thus happily ensuring, from the marketing
perspective of seed producers, a steady demand for their fresh
hybrid seeds).

Chapter Ten: Take Refuge

139 **the entire region lay under a continental ice sheet:** The same
principle that causes cold air to sink through a talus pile worked
at a much grander scale along the margins of the Laurentide Ice
Sheet. Air that became chilled and dense at the surface of the
mile-thick glacier spilled continuously over its edges, flowing
downward and creating what meteorologists call katabatic winds
that swept across the surrounding landscape at speeds often exceeding fifty-five miles per hour (ninety kilometers per hour)
during the winter months. See Bromwich et al. 2004 for one of
many models of ice age weather.

141 **the importance of refugia in the tropics:** Controversy about
tropical refugia revolves around their putative role in creating
new species. German ornithologist Jürgen Haffer famously proposed in 1969 that expansion and contraction of the Amazon
rainforest during the Pleistocene (or earlier) helped drive the
region's exceptional biodiversity. He argued that repeated con-
tractions into refugia provided high levels of reproductive isolation, allowing populations to diverge from one another at an
unusual rate. This paradigm held for decades, but it began to
unravel when pollen records cast doubt on the frequency and
extent of rainforest contraction, and when genetic studies failed
to support rapid Pleistocene speciation for most groups. There is
currently no consensus on why the Amazon is so diverse, other
than: it's complicated. For a recent review of the topic see Rocha
and Kaefer 2019.

141 **patterns still visible in the modern distribution of everything
from snails to primates:** Evidence for refugia in the Congo Basin
is far more consistent than in the Amazon, with measurable—
if not extreme—impacts on genetic diversity and speciation.
For diverse examples, see Wronski and Hausdorf 2008 for land
snails, Ntie et al. 2017 for forest antelopes, and Anthony et al.
2007 for gorillas.

142 **"the region of maximum sap flow":** Rapp et al. 2019, p. 187.

144 **new research has opened a window of hope:** "Ostracized, scrutinized, ignored." That's how Constance Millar summarized the initial scientific reaction to her and her colleagues' pika findings. Even though the findings were good news, the refugia conclusion ran counter to a long-established narrative that pikas were, in Millar's words, "temperate polar bears"—an iconic species on the verge of climate-driven extinction. In the long term that may still be true; no one knows how much time living in talus refugia will buy them. But Millar hopes her work can now help spread attention and research effort to other species at risk— things like alpine chipmunks that have no refuge to retreat to.

144 **cooler than their immediate surroundings during the summer:** The benefits of a talus slope for pikas include more than cool summer temperatures. Melting ice from deep inside trickles out downslope, helping to water the wet meadow vegetation they rely upon. And there is even an advantage in winter, when snowpack builds up on the rocks at the surface but leaves the interior airy and insulated from extreme cold. That's an important consideration since pikas don't hibernate; they prefer to spend the off-season awake, hunkered down in the dark, nibbling away on their accumulation of haystacks. See Millar and Westfall 2010 and related references.

Chapter Eleven: Pushing the Envelope

153 **"But who wants to be foretold the weather? . . .":** Jerome 1889, p. 36.

156 **mapped out the distinct bands of vegetation that grew along its flanks:** Though often interpreted literally, Von Humboldt's Chimborazo illustrations were intended as a general guide to vegetation patterns in the tropical Andes. Some of the species and communities listed were actually observed on other mountains, complicating recent attempts to use the diagram as a baseline for climate research on Chimborazo itself. See Moret et al. 2019.

160 **"Further details and examples will be given in a paper now in preparation.":** Holdridge 1947, p. 368.

160 field-tested his ideas throughout Costa Rica's many life zones and beyond: In an interesting twist, Leslie Holdridge received much of his research funding for work in Costa Rica from the United States Army. With the war escalating in Vietnam, the US military was suddenly quite interested in tropical landscapes, and whether a few simple climate variables could give an accurate prediction of conditions on the ground.

161 "conjecture.": Holdridge 1967, p. 79.

161 One prominent early climate paper invited readers to do just that: See Emanuel et al. (1985) for a fascinating direct link between Holdridge's work and the rising field of predictive modeling in climate change biology.

166 What emerged from the Audubon analysis: For the Audubon climate analysis, Wilsey and his team used an algorithm similar to Random Forest called a boosted regression tree, in combination with one memorably named maximum entropy. See Bateman et al. 2020.

166 they took the time to produce a flashy, interactive website: The Audubon Society's full "Survival by Degrees" report and maps can be explored online at www.audubon.org/climate/survival bydegrees.

168 but he cautioned against taking any model prediction too literally: There is a famous saying in statistics: "All models are wrong, but some are useful." Wilsey invoked this adage when he told me that the Audubon models don't need to get every detail right to provide informative insights about climate change and birds. "In spite of the uncertainties, there is power in comparing different scenarios," he explained. Warm the planet by 5.4 degrees Fahrenheit (3 degrees Celsius), the "business as usual" carbon future, and the models predict that nearly two-thirds of North America's birds will be moderately or highly threatened. But if warming can be limited to 2.7 degrees Fahrenheit (1.5 degrees Celsius), that figure drops to less than half. "What's clear is that taking action makes a difference," Wilsey concluded. "And that's a powerful message."

172 added carbon dioxide was having an effect too, giving many plants at least a temporary boost in growth rates: Because

plants use carbon dioxide during photosynthesis, adding more of it to the atmosphere should theoretically help them grow, but the relationship is far from straightforward. "Plants do like the added carbon dioxide," Kolka told me, "but something else, often nitrogen, quickly becomes limiting, so the boost is usually temporary."

Chapter Twelve: Surprise, Surprise

173 "... no matter how complete our past observations may have been.": McCrea 1963, p. 197.

173 and "the butterfly effect" was born: The origin of the term "butterfly effect" is often traced to a paper that Edward Lorenz presented at the 1972 meeting of the American Association for the Advancement of Science. But a colleague who arranged Lorenz's session later took credit for coming up with the now-famous title: "Predictability: Does the Flap of a Butterfly's Wings in Brazil Set Off a Tornado in Texas?" For his part, Lorenz couldn't recall precisely when he started referring to butterflies over seagulls, but it was a metaphor he had originally adapted from another meteorologist and had been using—in one form or another—for years. See Dooley 2009.

174 "Surprizes are foolish things . . .": Austen 2015, p. 183.

175 "... nearly four millions of birds on the wing at one time.": Beechey 1843, p. 46.

178 grew at precisely the same rate as they had on a traditional diet: At least two of the dovekies' preferred sea ice–dependent zooplankton varieties were unavailable to dovekies at the glacial/ocean water barrier. But the birds apparently made up for that lack by feeding on abundant small crustaceans called copepods. See Grémillet et al. 2015.

180 The stress of that physical damage appears to trigger a "now or never" reproductive surge: There is also the strong possibility that chemical cues in bumblebee saliva enhance this effect. When experimenters mimicked the bee damage with forceps and a razor, flowering times still advanced, but by a much smaller degree. See Pashalidou et al. 2020.

182 **the way pack rat urine crystallizes to form a hard, amber-like crust:** The technical term for ancient midden material is "amberat," a substance so hard it must be collected with a rock hammer and soaked for days in water to free its contents. Pack rat middens date back at least fifty thousand years, the limit of accurate carbon dating. Some middens may be far older.

184 **". . . create a product for which there is no market":** Lenz 2001, p. 61.

184 **their absence continues to reverberate biologically:** The full impacts of megafaunal extinctions are poorly understood and largely unstudied. One of the most intriguing possibilities comes from the Arctic, where some scientists believe the absence of grazing by mammoths, woolly rhinoceros, bison, and horses caused what had been a steppelike grassland to transform into the mossy tundra that dominates today. Experiments with horses, yaks, musk ox, and other ungulates at "Pleistocene Park," a private research reserve in Siberia, suggest that reintroducing missing herbivores could flip the system back to a grassland, potentially increasing carbon sequestration and slowing the loss of permafrost. See Macias-Fauria et al. 2020.

185 **a paltry dispersal rate of only six feet (two meters) per year:** Limited distance is not the only disadvantage of rodent dispersal. While Joshua tree seeds apparently passed through a sloth gut whole and unharmed, squirrels and rats gather and cache seeds with the intent of returning later to devour them. Successful dispersal depends on caches being lost or abandoned, but that may be rare. In one study, only 3 of 836 rodent-dispersed Joshua tree seeds survived to germination. See Vander Wall et al. 2006.

187 **human-driven trends have drastically altered ecosystems:** Climate change is a grave and overarching threat to biodiversity, a theme explored in Elizabeth Kolbert's excellent book, *The Sixth Extinction*. But before its impacts became so readily apparent, plants and animals were already highly threatened by other human activities—so much so that Kenyan conservationist and anthropologist Richard Leakey and journalist Roger Lewin wrote a book with the very same title two decades earlier. It made a similarly convincing argument that our species was triggering

the planet's sixth mass extinction event based entirely on things like habitat loss and overhunting. Modern climate change was not even mentioned.

Chapter Thirteen: That Was Then, This Is Now

189 **"The historian is a prophet facing backwards.":** Schlegel 1991, p. 27.

192 **shaggy, thick-necked ostriches with sledgehammer beaks:** Many taxonomists now lump the North American bird *Diatryma* with a similar genus from Europe called *Gastornis*. The largest varieties stood nearly seven feet (two meters) tall and were long considered fierce predators. But some experts think they were plant eaters, "gentle giants," in Renee Love's words, using their massive beaks to crush seeds, fruits, and woody vegetation.

192 **it probably has to do with how plants regulate water loss:** The precise relationship between leaf margins and temperature is one of those apparently simple patterns in nature that elude a precise explanation. Teeth do increase transpiration (water flow) potential, which could help plants maximize growth opportunities in cooler, seasonal environments. Conversely, lacking teeth may prevent dehydration in the tropics. Other research suggests it may be more straightforward: a direct function of efficiency in leaf growth under different conditions. See Wilf 1997.

193 **". . . gauging the general climatic conditions of the Cretaceous and Tertiary.":** Bailey and Sinnott 1916, p. 38.

194 **averaged fifteen to twenty-two degrees Fahrenheit (eight to twelve degrees Celsius) hotter:** Global temperatures during the Paleocene-Eocene Thermal Maximum averaged nine to sixteen degrees Fahrenheit (five to nine degrees Celsius) higher than current conditions. Love's data fall well above that range in part due to the site's latitude, but mostly because of its elevation. Sitting right at sea level, it's one of the few sites studied to date that supported a steamy, lowland rainforest.

194 **reaching Eocene levels by the middle of the next century:** Climate models that compare future scenarios with specific conditions from the past have identified the early Eocene as the best

historical analog for a world where current carbon emissions remain unabated. Restricting emissions to intermediate levels will put our future climate more along the lines of the mid-Pliocene, a warm period 3.3 million years ago when conditions were less extreme. See Burke et al. 2019.

195 **the asteroid incident that wiped out so many dinosaurs:** A competing theory attributes the demise of dinosaurs (and widespread simultaneous extinctions) to extended volcanic activity associated with the Deccan Traps in what is now India. But that too would have boiled down to a climate-driven process: short-term cooling episodes interspersed with carbon-dioxide-driven warming. Some experts now argue for a combined model—an asteroid strike and global winter decimating fauna and flora already stressed by climate instability.

195 **the disappearance of certain bottom-dwelling marine plankton:** Even before paleontologists understood how the climate had changed in the early Eocene, it was a time period known for extinctions in the foraminifera, a diverse group of tiny shell-making plankton that leave behind abundant fossils. The cause is assumed to be related to warmer, more acidic water, but it remains somewhat mysterious because it affected only a narrow group of mid- to deep-water bottom dwellers. See McInerney and Wing 2011.

197 **Subtle differences in water molecules:** Ratios of oxygen and hydrogen isotopes in water vary with changes in global temperature. The distinction has to do with how the rates of evaporation and precipitation for molecules containing rare, heavy isotopes differ as conditions warm.

198 **At least twenty-five times during the past 120,000 years:** Rapid temperature spikes during the past ice age are called Dansgaard–Oeschger events, after their Danish and Swiss co-discoverers. Two additional warming events occurred during the transition to the Holocene, the Bølling–Allerød and the warming at the end of the Younger Dryas. Both are far better known than the Dansgaard–Oeschger events and they're usually considered distinct, but some experts simply consider them the most recent examples of the same pattern.

200 **"the intersection of paleoecology, paleoclimatology, paleo-genomics . . . ":** Fordham et al. 2020, p. 1.

201 **the recovery and analysis of increasingly ancient DNA:** Another simple but telling insight comes from the overlap between paleogenomics and taxonomy, and involves the age of species. Plants and animals that evolved before the Pleistocene, for example, have already survived major climate upheavals and had the chance to develop and retain a wide array of adaptive traits. Newer, younger species groups lack that evolutionary history and may be at higher risk.

201 **"anatomically modern humans.":** Fordham et al. 2020, p. 3.

201 **Climate correlations are now being found for well-known events throughout history:** Each of these fascinating case studies is worthy of further reading. The volcano that influenced events in Rome, for example, erupted halfway around the globe on a tiny island in Alaska, but it spewed enough ash into the atmosphere to cool temperatures on the Italian peninsula by as much as thirteen degrees Fahrenheit (seven degrees Celsius) for two years. See McConnell et al. 2020, Pederson et al. 2014, Waldinger 2013.

204 **the 40 percent rise in armed conflicts since the year 2000:** Active armed conflicts around the globe have increased by nearly 40 percent since the turn of the twenty-first century. See Pettersson and Öberg 2020.

205 **measurable migration trends toward relative climate stability:** Not surprisingly, climate-driven migration is highest within and among countries dependent upon agriculture. But it is also strongly affected by wealth, peaking in middle-income nations where people have enough means to move, but not enough to pay for expensive local fixes. Migration rates drop off where it is unaffordable, or in places wealthy enough where people can remain comfortable in spite of changing conditions. It's an intriguing example of wealth as a form of social plasticity—those with more wealth can either move or purchase adaptations. See Hoffman et al. 2020.

205 **While writing this paragraph, I scanned the headlines:** My survey of the news didn't reveal anything to do with witch trials, but I did notice a story about another irrational search for scapegoats: the rising political influence of conspiracy theories. Social scientists associate increases in such "accusatory perceptions" with parallel rises in conflict, stress, and traumatic events. Notably, a number of current conspiracy theories have to do with the origins and veracity of climate change itself.

Conclusion: Everything You Can

207 **"Strong reasons make strong actions.":** *King John*, Act III, Scene 4; Bevington 1980, p. 470.

211 **"And because I cannot do everything . . .":** Fairfield 1890, p. 114.

Bibliography

Anderson, J. T., and Z. J. Gezon. 2014. Plasticity in functional traits in the context of climate change: a case study of the subalpine forb *Boechera stricta* (Brassicaceae). *Global Change Biology* 21: 1689–1703.

Anthony, N. M., M. Johnson-Bawe, K. Jeffery, S. L. Clifford, et al. 2007. The role of Pleistocene refugia and rivers in shaping gorilla genetic diversity in central Africa. *Proceedings of the National Academy of Sciences* 104: 20432–20436.

Aronson, R. B., K. E. Smith, S. C. Vos, J. B. McClintock, et al. 2015. No barrier to emergence of bathyal king crabs on the Antarctic Shelf. *Proceedings of the National Academy of Sciences* 112: 12997–13002.

Arrhenius, S. 1908. *Worlds in the Making: The Evolution of the Universe*. New York: Harper and Brothers.

Aubret, F., and R. Shine. 2010. Thermal plasticity in young snakes: how will climate change affect the thermoregulatory tactics of ectotherms? *The Journal of Experimental Biology* 213: 242–248.

Austen, J. 2015 (1816). *Emma*. 200th Anniversary Annotated Edition. New York: Penguin.

Bailey, I. W., and E. W. Sinnott. 1916. The climatic distribution of certain types of angiosperm leaves. *American Journal of Botany* 3: 24–39.

Barnosky, A. 2014. *Dodging Extinction: Power, Food, Money, and the Future of Life on Earth*. Oakland: University of California Press.

Barnosky, A. D., P. L. Koch, R. S. Feranec, S. L. Wing, et al. 2004. Assessing the causes of late Pleistocene extinctions on the continents. *Science* 306: 70–75.

Bateman, B. L., L. Taylor, C. Wilsey, J. Wu, et al. 2020. Risk to North American birds from climate change–related threats. *Conservation Science and Practice* 2. DOI: 10.1111/csp2.243.

Bateman, B. L., C. Wilsey, L. Taylor, J. Wu, et al. 2020. North American birds require mitigation and adaptation to reduce vulnerability to climate change. *Conservation Science and Practice* 2. DOI: 10.1111/csp2.242.

Bates, A. E., B. J. Hilton, and C. D. G. Harley. 2009. Effects of temperature, season and locality on wasting disease in the keystone predatory sea star *Pisaster ochraceus*. *Diseases of Aquatic Organisms* 86: 245–251.

Bateson, P., D. Barker, T. Clutton-Brock, D. Deb, et al. 2004. Developmental plasticity and human health. *Nature* 430: 419–421.

Becker, M., N. Gruenheit, M. Steel, C. Voelckel, et al. 2013. Hybridization may facilitate in situ survival of endemic species through periods of climate change. *Nature Climate Change* 3: 1039–1043.

Bednaršek, N., R. A. Feely, J. C. P. Reum, B. Peterson, et al. 2014. *Limacina helicina* shell dissolution as an indicator of declining habitat suitability owing to ocean acidification in the California Current Ecosystem. *Proceedings of the Royal Society B* 281: 20140123.

Beechey, F. W. 1843. *A Voyage of Discovery Towards the North Pole*. London: Richard Bentley.

Bellard, C., W. Thuiller, B. Leroy, P. Genovesi, et al. 2013. Will climate change promote future invasions? *Global Change Biology* 12: 3740–3748.

Bevington, D., ed. 1980. *The Complete Works of Shakespeare*. Glenview, IL: Scott, Foresman and Company.

Blom, P. 2017. *Nature's Mutiny*. New York: Liveright Publishing Company.

Bordier, C., H. Dechatre, S. Suchail, M. Peruzzi, et al. 2017. Colony adaptive response to simulated heat waves and consequences at the individual level in honeybees (*Apis mellifera*). *Scientific Reports* 7: 3760.

Botkin, D. B., H. Saxe, M. B. Araújo, R. Betts, et al. 2007. Forecasting the effects of global warming on biodiversity. *BioScience* 57: 227–236.

Botta, F., D. Dahl-Jensen, C. Rahbek, A. Svensson, et al. 2019. Abrupt change in climate and biotic systems. *Current Biology* 29: R1045–R1054.

Boutin, S., and J. E. Lane. 2014. Climate change and mammals: evolutionary versus plastic responses. *Evolutionary Applications* 7: 29–41.

Brakefield, P. M., and P. W. de Jong. 2011. A steep cline in ladybird melanism has decayed over 25 years: a genetic response to climate change? *Heredity* 107: 574–578.

Breedlovestrout, R. L. 2011. "Paleofloristic Studies in the Paleogene Chuckanut Basin, Western Washington, USA." PhD dissertation. Moscow: University of Idaho, 952 pp.

Breedlovestrout, R. L., B. J. Evraets, and J. T. Parrish. 2013. New Paleogene paleoclimate analysis of western Washington using physiognomic characteristics from fossil leaves. *Palaeogeography, Palaeoclimatology, Palaeoecology* 392: 22–40.

Bromwich, D. H., E. R. Toracinta, H. Wei, R. J. Oglesby, et al. 2004. Polar MM5 simulations of the winter climate of the Laurentide Ice Sheet at the LGM. *Journal of Climate* 17: 3415–3433.

Brooker, R. M., S. J. Brandl, and D. L. Dixson. 2016. Cryptic effects of habitat declines: coral-associated fishes avoid coral-seaweed interactions due to visual and chemical cues. *Scientific Reports* 6: 18842.

Burke, K. D., J. W. Williams, M. A. Chandler, A. M. Haywood, et al. 2018. Pliocene and Eocene provide best analogs for near-future climates. *Proceedings of the National Academy of Sciences* 115: 13288–13293.

Burnet, J. 1892. *Early Greek Philosophy*. London: Adam and Charles Black.

Candolin, U., T. Salesto, and M. Evers. 2007. Changed environmental conditions weaken sexual selection in sticklebacks. *Journal of Evolutionary Biology* 20: 233–239.

Carlson, S. M. 2017. Synchronous timing of food resources triggers bears to switch from salmon to berries. *Proceedings of the National Academy of Sciences* 114: 10309–10311.

Caruso, N. M., M. W. Sears, D. C. Adams, and K. R. Lips. 2014. Widespread rapid reductions in body size of adult salamanders in response to climate change. *Global Change Biology* 20: 1751–1759.

Carver, T. N. 1915. *Essays in Social Justice*. Cambridge, MA: Harvard University Press.

Chan-McLeod, A. C. A. 2006. A review and synthesis of the effects of unsalvaged mountain-pine-beetle-attacked stands on wildlife and implications for forest management. *BC Journal of Ecosystems and Management* 7: 119–132.

Chen, I., J. K. Hill, R. Ohlemüller, D. B. Roy, et al. 2011. Rapid range shifts of species associated with high levels of climate warming. *Science* 333: 1024–1026.

Christie, K. S., and T. E. Reimchen. 2008. Presence of salmon increases passerine density on Pacific Northwest streams. *The Auk* 125: 51–59.

Clairbaux, M., J. Fort, P. Mathewson, W. Porter, H. Strøm, et al. 2019. Climate change could overturn bird migration: transarctic flights and high-latitude residency in a sea ice free Arctic. *Scientific Reports* 9: 1–13.

Clark, J. S., C. Fastie, G. Hurtt, S. T. Jackson, et al. 1998. Reid's paradox of rapid plant migration: dispersal theory and interpretation of paleoecological records. *BioScience* 48: 13–24.

Cleese, J., E. Idle, G. Chapman, T. Jones, et al. 1974. *Monty Python and the Holy Grail Screenplay*. London: Methuen.

Cole, K. L., K. Ironside, J. Eischeid, G. Garfin, et al. 2011. Past and ongoing shifts in Joshua tree distribution support future modeled range contraction. *Ecological Applications* 21: 137–149.

Cooke, B. J., and A. J. Carroll. 2017. Predicting the risk of mountain pine beetle spread to eastern pine forests: considering uncertainty in uncertain times. *Forest Ecology and Management* 396: 11–25.

Corlett, R. T., and D. A. Westcott. 2013. Will plant movements keep up with climate change? *Trends in Ecology & Evolution* 28: 482–488.

Crawford, E. 1996. *Arrhenius: From Ionic Theory to the Greenhouse Effect*. Canton, MA: Science History Publications.

Crimmins S., S. Dobrowski, J. Greenberg, J. Abatzoglou, et al. 2011. Changes in climatic water balance drive downhill shifts in plant species' optimum elevations. *Science* 331: 324–327.

Cronin, T. M. 2010. *Paleoclimates*. New York: Columbia University Press.

Crozier, L. G., and J. A. Hutchings. 2014. Plastic and evolutionary responses to climate change in fish. *Evolutionary Applications* 7: 68–87.

Cudmore, T. J., N. Björklund, A. L. Carroll, and S. Lindgren. 2010. Climate change and range expansion of an aggressive bark beetle: evidence of higher beetle reproduction in naïve host tree populations. *Journal of Applied Ecology* 47: 1036–1043.

da Rocha, G. D., and I. L. Kaefer. 2019. What has become of the refugia hypothesis to explain biological diversity in Amazonia? *Ecology and Evolution* 9: 4302–4309.

Darwin, C. 2004. *The Voyage of the Beagle* (1909 text). Washington, DC: National Geographic Adventure Classics.

Darwin, C. 2008. *On the Origin of Species: The Illustrated Edition* (1859 text). New York: Sterling.

Deacy, W. W., J. B. Armstrong, W. B. Leacock, C. T. Robbins, et al. 2017. Phenological synchronization disrupts trophic interactions between Kodiak brown bears and salmon. *Proceedings of the National Academy of Sciences* 114: 10432–10437.

Deacy, W., W. Leacock, J. B. Armstrong, and J. A. Stanford. 2016. Kodiak brown bears surf the salmon red wave: direct evidence from GPS collared individuals. *Ecology* 97: 1091–1098.

Dessler, A. 2016. *Introduction to Modern Climate Change*. New York: Cambridge University Press.

di Lampedusa, G. 1960. *The Leopard*. New York: Pantheon Books.

Donihue, C. M., A. Herrel, A. C. Fabre, A. Kamath, et al. 2018. Hurricane-induced selection on the morphology of an island lizard. *Nature* 560: 88–91.

Dooley, K. J. 2009. The butterfly effect of the "butterfly effect." *Nonlinear Dynamics, Psychology, and Life Sciences* 13: 279–288.

Draper, A. M., and M. Weissburg. 2019. Impacts of global warming and elevated CO_2 on sensory behavior in predator-prey interactions: a review and synthesis. *Frontiers in Ecology and Evolution* 7: 72–91.

Edworthy, A. B., M. C. Drever, and K. Martin. 2011. Woodpeckers increase in abundance but maintain fecundity in response to an outbreak of mountain pine bark beetles. *Forest Ecology and Management* 261: 203–210.

Eisenlord, M. E., M. L. Groner, R. M. Yoshioka, J. Elliott, et al. 2016. Ochre star mortality during the 2014 wasting disease epizootic: role of population size structure and temperature. *Philosophical Transactions of the Royal Society B: Biological Sciences* 371: 20150212. DOI: 1098/rstb.2015.0212.

Ekmarck, D. 1781. On the Migration of Birds. In F. J. Brand, transl., *Select Dissertations from the Amoenitates Academicae* 215–263. London: G. Robinson, Bookseller.

Eldredge, N., and S. J. Gould. 1972. "Punctuated Equilibria: An Alternative to Phyletic Gradualism." In T. J. M. Schopf, ed., *Models in Paleobiology*, 82–115. San Francisco: Freeman, Cooper & Co.

Ellwood, E. R., J. M. Diez, I. Ibánez, R. B. Primack, et al. 2012. Disentangling the paradox of insect phenology: are temporal trends reflecting the response to warming? *Oecologia* 168: 1161–1171.

Ellwood, E. R., S. A. Temple, R. B. Primack, N. L. Bradley, et al. 2013. Recordbreaking early flowering in the eastern United States. *PLoS One* 8: e53788.

Emanuel, W. R., H. H. Shugart, and M. P. Stevenson. 1985. Climatic change and the broad-scale distribution of terrestrial ecosystem complexes. *Climatic Change* 7: 29–43.

Erlenbach, J. A., K. D. Rode, D. Raubenheimer, and C. T. Robbins. 2014. Macronutrient optimization and energy maximization determine diets of brown bears. *Journal of Mammalogy* 95: 160–168.

Evans, S. R., and L. Gustafsson. 2017. Climate change upends selection on ornamentation in a wild bird. *Nature Ecology & Evolution* 1: 1–5.

Fagan, B. 2000. *The Little Ice Age: How Climate Made History*. New York: Basic Books.

Fagen, J. M., and R. Fagen. 1994. Bear-human interactions at Pack Creek, Alaska. *International Conference on Bear Research and Management* 9: 109–114.

Fairfield, A. H., ed. 1890. *Starting Points: How to Make a Good Beginning*. Chicago: Young Men's Era Publishing Company.

Fei, S., J. M. Desprez, K. M. Potter, I. Jo, et al. 2017. Divergence of species responses to climate change. *Science Advances* 3: e1603055.

Fordham, D. A., S. T. Jackson, S. C. Brown, B. Huntley, et al. 2020. Using paleo-archives to safeguard biodiversity under climate change. *Science* 369: eabc5654. DOI: 10.1126/science.abc5654.

Foster, D. R., and T. M. Zebryk. 1993. Long-term vegetation dynamics and disturbance history of a *Tsuga*-dominated forest in New England. *Ecology* 74: 982–998.

Franks, S. J., J. J. Webber, and S. N. Aitken. 2014. Evolutionary and plastic responses to climate change in terrestrial plant populations. *Evolutionary Applications* 7: 123–139.

Freeman, B. G., and A. M. C. Freeman. 2014. Rapid upslope shifts in New Guinean birds illustrate strong distributional responses of tropical montane species to global warming. *Proceedings of the National Academy of Sciences* 111: 4490–4494.

Freeman B. G., J. A. Lee-Yaw, J. Sunday, and A. L. Hargreaves. 2017. Expanding, shifting and shrinking: the impact of global warming on species' elevational distributions. *Global Ecology and Biogeography* 27: 1268–1276.

Freeman, B. G., M. N. Scholer, V. Ruiz-Gutierrez, and J. W. Fitzpatrick. 2018. Climate change causes upslope shifts and mountaintop extirpations in a tropical bird community. *Proceedings of the National Academy of Sciences* 115: 11982–11987.

Fritz, A. 2017. This city in Alaska is warming so fast, algorithms removed the data because it seemed unreal. *The Washington Post*, December 12, 2017. Archived at www.washingtonpost.com. Accessed March 20, 2019.

Gallinat, A. S., R. B. Primack, and D. L. Wagner. 2015. Autumn, the neglected season in climate change research. *Trends in Ecology and Evolution* 30: 169–176.

Gardner, J., C. Manno, D. C. Bakker, V. L. Peck, et al. 2018. Southern Ocean pteropods at risk from ocean warming and acidification. *Marine Biology* 165. DOI: 10.1007/s00227-017-3261-3.

Gienapp, P., C. Teplitsky, J. S. Alho, J. A. Mills, et al. 2008. Climate change and evolution: disentangling environmental and genetic responses. *Molecular Ecology* 17: 167–178.

Gould, S. J. 2007. *Punctuated Equilibrium*. Cambridge, MA: The Belknap Press of Harvard University Press.

Grant, P. R., B. R. Grant, R. B. Huey, M. T. Johnson, et al. 2017. Evolution caused by extreme events. *Philosophical Transactions of the Royal Society B: Biological Sciences* 372: 20160146. DOI: 10.1098/rstb.2016.014.

Greiser, C., J. Ehrlén, E. Meineri, and K. Hylander. 2019. Hiding from the climate: characterizing microrefugia for boreal forest understory species. *Global Change Biology* 26: 471–483.

Grémillet, D., J. Fort, F. Amélieneau, E. Zakharova, et al. 2015. Arctic warming: nonlinear impacts of sea-ice and glacier melt on seabird foraging. *Global Change Biology* 21: 1116–1123.

Hannah, L. 2015. *Climate Change Biology.* 2nd Edition. London: Academic Press.

Hanson, T., W. Newmark, and W. Stanley. 2007. Forest fragmentation and predation on artificial nests in the Usambara Mountains, Tanzania. *African Journal of Ecology* 45: 499–507.

Harvell, C. D., D. Montecino-Latorre, J. M. Caldwell, J. M. Burt, et al. 2019. Disease epidemic and a marine heat wave are associated with the continental-scale collapse of a pivotal predator (*Pycnopodia helianthoides*). *Science Advances* 5: eaau7042. DOI: 0.1126/sciadv.aau7042.

Harvell, D. 2019. *Ocean Outbreak: Confronting the Tide of Marine Disease.* Oakland: University of California Press.

Hassal, C., S. Keat, D. J. Thompson, and P. C. Watts. 2014. Bergmann's rule is maintained during a rapid range expansion in a damselfly. *Global Change Biology* 20: 475–482.

Heberling, J. M., M. McDonough, J. D. Fridley, S. Kalisz, et al. 2019. Phenological mismatch with trees reduces wildflower carbon budgets. *Ecology Letters* 22: 616–623.

Hegarty, M. J., and S. J. Hiscock. 2005. Hybrid speciation in plants: new insights from molecular studies. *New Phytologist* 165: 411–423.

Heller, J. L. 1983. Notes on the titulature of Linnaean dissertations. *Taxon* 32: 218–252.

Hendry, A. P., K. M. Gotanda, and E. I. Svensson. 2017. Human influences on evolution, and the ecological and societal consequences. *Philosophical Transactions of the Royal Society B* 372: 20160028.

Herbert, S. 2005. *Charles Darwin, Geologist.* Ithaca, NY: Cornell University Press.

Hewson, I., J. B. Button, B. M. Gudenkauf, B. Miner, et al. 2014. Densovirus associated with sea-star wasting disease and mass mortality. *Proceedings of the National Academy of Sciences* 111: 17278–17283.

Hilborn, R. C. 2004. Sea gulls, butterflies, and grasshoppers: a brief history of the butterfly effect in nonlinear dynamics. *American Journal of Physics* 72: 425–427.

Hill, J. K., C. D. Thomas, and D. S. Blakely. 1999. Evolution of flight morphology in a butterfly that has recently expanded its geographic range. *Oecologia* 121: 165–170.

Hocking, M. D., and T. E. Reimchen. 2002. Salmon-derived nitrogen in terrestrial invertebrates from coniferous forests of the Pacific Northwest. *BMC Ecology* 2: 4.

Hoffmann, R., A. Dimitrova, R. Muttarak, J. Crespo Cuaresma, et al. 2020. A meta-analysis of country-level studies on environmental change and migration. *Nature Climate Change* 10. DOI: 10.1038 /s41558-020-0898-6.

Holdridge, L. R. 1947. Determination of world plant formations from simple climatic data. *Science* 105: 367–368.

Holdridge, L. R. 1967. *Life Zone Ecology*. San Jose, Costa Rica: Tropical Science Center.

Honey-Marie, C., A. L. Carroll, and B. H. Aukema. 2012. Breach of the northern Rocky Mountain geoclimatic barrier: initiation of range expansion by the mountain pine beetle. *Journal of Biogeography* 39: 1112–1123.

Hort, A., transl. 1938. *The Critica Botanica of Linnaeus*. London: The Ray Society.

Hoving, H.-J. T., W. F. Gilly, U. Markaida, K. J. Benoit-Bird, et al. 2013. Extreme plasticity in life-history strategy allows a migratory predator (jumbo squid) to cope with a changing climate. *Global Change Biology* 19: 2089–2103.

Huey, R. B., J. B. Losos, and C. Moritz. 2010. Are lizards toast? *Science* 328: 832–833.

Hulme, M. 2009. On the origin of "the greenhouse effect": John Tyndall's 1859 interrogation of nature. *Weather* 64: 121–123.

Hutton, J. 1788. Theory of the earth. *Transactions of the Royal Society of Edinburgh* 1: 209.

Isaak, D. J., M. K. Young, C. H. Luce, S. W. Hostetler, et al. 2016. Slow climate velocities of mountain streams portend their role as refugia for cold-water biodiversity. *Proceedings of the National Academy of Sciences* 113: 4374–4379.

Jefferson, T. 1803. Jefferson's instructions to Meriwether Lewis. Letter dated June 20, 1803. Archived at www.monticello.org. Accessed October 31, 2018.

Johnson, C. R., S. C. Banks, N. S. Barrett, F. Cazassus, et al. 2011. Climate change cascades: shifts in oceanography, species' ranges and subtidal marine community dynamics in eastern Tasmania. *Journal of Experimental Marine Biology and Ecology* 400: 17–32.

Johnson, S. 2008. *The Invention of Air*. New York: Riverhead Books.

Johnson, W. C., and C. S. Adkisson. 1986. Airlifting the oaks. *Natural History* 95: 40–47.

Johnson, W. C., and T. Webb III. 1989. The role of blue jays (*Cyanocitta cristata* L.) in the postglacial dispersal of fagaceous trees in eastern North America. *Journal of Biogeography* 16: 561–571.

Johnson-Groh, C., and D. Farrar. 1985. Flora and phytogeographical history of Ledges State Park, Boone County, Iowa. *Proceedings of the Iowa Academy of Science* 92: 137–143.

Jost, J. T. 2015. Resistance to change: a social psychological perspective. *Social Research* 82: 607–636.

Karell, P., K. Ahola, T. Karstinen, J. Valkama, et al. 2011. Climate change drives microevolution in a wild bird. *Nature Communications* 2: 1–7.

Keith, S. A., A. H. Baird, J. P. A. Hobbs, E. S. Woolsey, et al. 2018. Synchronous behavioural shifts in reef fishes linked to mass coral bleaching. *Nature Climate Change* 8: 986–991.

Kirchman, J. J., and K. J. Schneider. 2014. Range expansion and the breakdown of Bergmann's Rule in red-bellied woodpeckers (*Melanerpes carolinus*). *The Wilson Journal of Ornithology* 126: 236–248.

Koch, A., C. Brierley, M. M. Maslin, and S. L. Lewis. 2019. Earth system impacts of the European arrival and Great Dying in the Americas after 1492. *Quaternary Science Reviews* 207: 13–36.

Kolbert, E. 2014. *The Sixth Extinction: An Unnatural History*. New York: Henry Holt.

Kooiman, M., and J. Amash. 2011. *The Quality Companion*. Raleigh, NC: TwoMorrows Publishing.

Körner, C., and E. Spehn. 2019. A Humboldtian view of mountains. *Science* 365: 1061.

Kovach, R. P., B. K. Hand, P. A. Hohenlohe, T. F. Cosart, et al. 2016. Vive la résistance: genome-wide selection against introduced alleles in invasive hybrid zones. *Proceedings of the Royal Society B: Biological Sciences* 283: 20161380.

Kutschera, U. 2003. A comparative analysis of the Darwin-Wallace papers and the development of the concept of natural selection. *Theory in Biosciences* 122: 343–359.

Kuzawa, C. W., and J. M. Bragg. 2012. Plasticity in human life history strategy: implications for contemporary human variation and the evolution of genus *Homo*. *Current Anthropology* 53: S369–S382.

LaBar, T., and C. Adami. 2017. Evolution of drift robustness in small populations. *Nature Communications* 8: 1–12.

La Sorte, F., and F. Thompson. 2007. Poleward shifts in winter ranges of North American birds. *Ecology* 88: 1803–1812.

Lenoir, J., J. C. Gégout, A. Guisan, P. Vittoz, et al. 2010. Going against the flow: potential mechanisms for unexpected downslope range shifts in a warming climate. *Ecography* 33: 295–303.

Lenz, L. W. 2001. Seed dispersal in *Yucca brevifolia* (Agavaceae)—present and past, with consideration of the future of the species. *Aliso: A Journal of Systematic and Evolutionary Botany* 20: 61–74.

Le Row, C. B. 1887. *English as She is Taught: Genuine Answers to Examination Questions in our Public Schools*. New York: Cassell and Company.

Liao, W., C. T. Atkinson, D. A. LaPointe, and M. D. Samuel. 2017. Mitigating future avian malaria threats to Hawaiian forest birds from climate change. *PLoS One* 12: e0168880. https://doi.org/10.1371/journal.pone.0168880.

Lindgren, B. S., and K. F. Raffa. 2013. Evolution of tree killing in bark beetles (Coleoptera: Curculionidae): trade-offs between the maddening crowds and a sticky situation. *The Canadian Entomologist* 145: 471–495.

Ling, S. D., C. R. Johnson, K. Ridgeway, A. J. Hobday, et al. 2009. Climate-driven range extension of a sea urchin: inferring future trends by analysis of recent population dynamics. *Global Change Biology* 15: 719–731.

Little, A. G., D. N. Fisher, T. W. Schoener, and J. N. Pruitt. 2019. Population differences in aggression are shaped by tropical cyclone-induced selection. *Nature Ecology and Evolution* 3: 1294–1297.

Lorenz, E. N. 1963. The predictability of hydrodynamic flow. *Transactions of the New York Academy of Sciences*, Series II 25: 409–432.

Lourenço, C. R., G. I. Zardi, C. D. McQuaid, E. A. Serrao, et al. 2016. Upwelling areas as climate change refugia for the distribution and genetic diversity of a marine macroalga. *Journal of Biogeography* 43: 1595–1607.

Mabey, R. 1986. *Gilbert White: A Biography of the Author of "The Natural History of Selborne."* London: Century Hutchinson Ltd.

Macias-Fauria, M., P. Jepson, N. Zimov, and Y. Malhi. 2020. Pleistocene Arctic megafaunal ecological engineering as a natural climate solution? *Philosophical Transactions of the Royal Society B* 375: 20190122. DOI: 10.1098/rstb.2019.0122.

Mackay, C. 1859. *The Collected Songs of Charles Mackay.* London: G. Routledge and Co.

Mackey, B., S. Berry, S. Hugh, S. Ferrier, et al. 2012. Ecosystem greenspots: identifying potential drought, fire, and climate-change micro-refuges. *Ecological Applications* 22: 1852–1864.

Marshall, G. 2014. *Don't Even Think About It: Why Our Brains Are Wired to Ignore Climate Change.* New York: Bloomsbury.

Martin, T.-H. 1868. *Galilée: Les Droits de la Science et la Méthode des Sciences Physiques.* Paris: Didier et Cie.

Mayhew, P. J., G. B. Jenkins, and T. G. Benton. 2008. A long-term association between global temperature and biodiversity, origination and extinction in the fossil record. *Proceedings of the Royal Society B: Biological Sciences* 275: 47–53.

McConnell, J. R., M. Sigl, G. Plunkett, A. Burke, et al. 2020. Extreme climate after massive eruption of Alaska's Okmok volcano in 43 BCE and effects on the late Roman Republic and Ptolemaic Kingdom. *Proceedings of the National Academy of Sciences* 117: 15443–15449.

McCrea, W. H. 1963. Cosmology, a brief review. *Quarterly Journal of the Royal Astronomical Society* 4: 185–202.

McEvoy, B. P., and P. M. Visscher. 2009. Genetics of human height. *Economics & Human Biology* 7: 294–306.

McInerney, F. A., and S. L. Wing. 2011. The Paleocene-Eocene Thermal Maximum: a perturbation of carbon cycle, climate, and biosphere with implications for the future. *Annual Review of Earth and Planetary Sciences* 39: 489–516.

Merilä, J., and A. P. Hendry. 2014. Climate change, adaptation, and phenotypic plasticity: the problem and the evidence. *Evolutionary Applications* 7: 1–14.

Millar, C. I., D. L. Delany, K. A. Hersey, M. R. Jeffress, et al. 2018. Distribution, climatic relationships, and status of American pikas (*Ochotona princeps*) in the Great Basin, USA. *Arctic, Antarctic, and Alpine Research* 50: p.e1436296.

Millar, C. I., and R. D. Westfall. 2010. Distribution and climatic relationships of the American pika (*Ochotona princeps*) in the Sierra Nevada and western Great Basin, USA: periglacial landforms as refugia in warming climates. *Arctic, Antarctic, and Alpine Research* 42: 76–88.

Millar, C. I., R. D. Westfall, and D. L. Delany. 2014. Thermal regimes and snowpack relations of periglacial talus slopes, Sierra Nevada, California, USA. *Arctic, Antarctic, and Alpine Research* 46: 483–504.

Millar, C. I., R. D. Westfall, and D. L. Delany. 2016. Thermal components of American pika habitat—how does a small lagomorph encounter climate? *Arctic, Antarctic, and Alpine Research* 48: 327–343.

Miller, M. 1974. *Plain Speaking: An Oral Biography of Harry S. Truman*. New York: G. P. Putnam's Sons.

Miller-Rushing, A. J., and R. B. Primack. 2008. Global warming and flowering times in Thoreau's Concord: a community perspective. *Ecology* 89: 332–341.

Mitton, J. B., and S. M. Ferrenberg. 2012. Mountain pine beetle develops an unprecedented summer generation in response to climate warming. *The American Naturalist* 179: E163–E171.

Morelli, T. L., C. Daly, S. Z. Dobrowski, D. M. Dulen, et al. 2016. Managing climate change refugia for climate adaptation. *PLoS One* 11: e0159909.

Moret, P., P. Muriel, R. Jaramillo, and O. Dangles. 2019. Humboldt's Tableau Physique revisited. *Proceedings of the National Academy of Sciences* 116: 12889–12894.

Moritz, C., and R. Agudo. 2013. The future of species under climate change: resilience or decline? *Science* 341: 505–508.

Muhlfeld, C. C., R. P. Kovach, R. Al-Chokhachy, S. J. Amish, et al. 2017. Legacy introductions and climatic variation explain spatio-temporal patterns of invasive hybridization in a native trout. *Global Change Biology* 23: 4663–4674.

Muhlfeld, C. C., R. P. Kovach, L. A. Jones, R. Al-Chokhachy, et al. 2014. Invasive hybridization in a threatened species is accelerated by climate change. *Nature Climate Change* 4: 620–624.

Newmark, W. D., and T. R. Stanley. 2011. Habitat fragmentation reduces nest survival in an Afrotropical bird community in a biodiversity hotspot. *Proceedings of the National Academy of Sciences* 108: 11488–11493.

Nogués-Bravo, D., F. Rodríguez-Sánchez, L. Orsini, E. de Boer, et al. 2018. Cracking the code of biodiversity responses to past climate change. *Trends in Ecology & Evolution* 33: 765–776.

Ntie, S., A. R. Davis, K. Hils, P. Mickala, et al. 2017. Evaluating the role of Pleistocene refugia, rivers and environmental variation in the diversification of central African duikers (genera *Cephalophus* and *Philantomba*). *BMC Evolutionary Biology* 17: 212. DOI: 10.1186/s12862-017-1054-4.

Ovadiah, A., and S. Mucznik. 2017. Myth and reality in the battle between the Pygmies and the cranes in the Greek and Roman worlds. *Gerión* 35: 151–166.

Parker, G. 2017. *Global Crisis: War, Climate Change and Catastrophe in the Seventeenth Century*. New Haven, CT: Yale University Press.

Parmesan, C. 2006. Ecological and evolutionary responses to recent climate change. *Annual Review of Ecology, Evolution, and Systematics* 37: 637–669.

Parmesan, C., and M. E. Hanley. 2015. Plants and climate change: complexities and surprises. *Annals of Botany* 116: 849–864.

Pashalidou, F. G., H. Lambert, T. Peybernes, M. C. Mescher, et al. 2020. Bumble bees damage plant leaves and accelerate flower production when pollen is scarce. *Science* 368: 881–884.

Pateman, R. M., J. K. Hill, D. B. Roy, R. Fox, et al. 2012. Temperature-dependent alterations in host use drive rapid range expansion in a butterfly. *Science* 336: 1028–1030.

Peck, V. L., R. L. Oakes, E. M. Harper, C. Manno, et al. 2018. Pteropods counter mechanical damage and dissolution through extensive shell repair. *Nature Communications* 9. DOI: 10.1038 /s41467-017-02692-w.

Peck, V. L., G. A. Tarling, C. Manno, E. M. Harper, et al. 2016. Outer organic layer and internal repair mechanism protects pteropod *Limacina helicina* from ocean acidification. *Deep Sea Research, Part II: Topical Studies in Oceanography* 127: 41–52.

Pecl, G. T., M. B. Araújo, J. D. Bell, J. Blanchard, et al. 2017. Biodiversity redistribution under climate change: impacts on ecosystems and human well-being. *Science* 355: eaai9214. DOI: 10.1126 /science.aai9214.

Pederson, N., A. E. Hessl, N. Baatarbileg, K. J. Anchukaitis, et al. 2014. Pluvials, droughts, the Mongol Empire, and modern Mongolia. *Proceedings of the National Academy of Sciences* 111: 4375–4379.

Petit, J. R., J. Jouzel, D. Raynaud, N. I. Barkov, et al. 1999. Climate and atmospheric history of the past 420,000 years from the Vostok ice core, Antarctica. *Nature* 399: 429–436.

Pettersson, T., and M. Öberg. 2020. Organized violence, 1989–2019. *Journal of Peace Research* 57: 597–613.

Pfister, C. A., R. T. Paine, and J. T. Wootton. 2016. The iconic keystone predator has a pathogen. *Frontiers in Ecology and the Environment* 14: 285–286.

Porfirio, L. L., R. M. Harris, E. C. Lefroy, S. Hugh, et al. 2014. Improving the use of species distribution models in conservation planning and management under climate change. *PLoS One* 9: e113749.

Prevey, J. S. 2020. Climate change: flowering time may be shifting in surprising ways. *Current Biology* 30: R112–R114.

Priestley, J. 1781. *Experiments and Observations on Different Kinds of Air*. London: J. Johnson.

Primack, R. B. 2014. *Walden Warming: Climate Change Comes to Thoreau's Woods*. Chicago: The University of Chicago Press.

Primack, R. B., and A. S. Gallinat. 2016. Spring budburst in a changing climate. *American Scientist* 104: 102–109.

Prum, R. O. 2017. *The Evolution of Beauty: How Darwin's Forgotten Theory of Mate Choice Shapes the Animal World*. New York: Doubleday.

Rapp, J. M., D. A. Lutz, R. D. Huish, B. Dufour, et al. 2019. Finding the sweet spot: shifting optimal climate for maple syrup production in North America. *Forest Ecology and Management* 448: 187–197.

Raup, D. M. 1994. The role of extinction in evolution. *Proceedings of the National Academy of Sciences* 91: 6758–6763.

Real, D., A. G. McAdam, S. Boutin, and D. Berteaux. 2003. Genetic and plastic responses of a northern mammal to climate change. *Proceedings of the Royal Society B* 270: 591–596.

Reed, T. E., V. Grotan, S. Jenouvrier, B. Saether, et al. 2013. Population growth in a wild bird is buffered against phenological mismatch. *Science* 340: 488–491.

Reid, C. 1899. *The Origin of the British Flora*. London: Dulau and Company.

Robbirt, K. M., D. L. Roberts, M. J. Hutchings, and A. J. Davy. 2014. Potential disruption of pollination in a sexually deceptive orchid by climatic change. *Current Biology* 24: 845–849.

Rosenberger, D. W., R. C. Venette, M. P. Maddox, and B. H. Aukema. 2017. Colonization behaviors of mountain pine beetle on novel hosts: implications for range expansion into northeastern North America. *PloS One* 12: e0176269.

Safranyik, L., and B. Wilson, eds. 2006. *The Mountain Pine Beetle: A Synthesis of Biology, Management and Impacts on Lodgepole Pine*. Victoria, BC: Canadian Forest Service.

Saintilan, N., N. Wilson, K. Rogers, A. Rajkaran, et al. 2014. Mangrove expansion and salt marsh decline at mangrove poleward limits. *Global Change Biology* 20: 147–157.

Sanford, E., J. L. Sones, M. García-Reyes, J. H. Goddard, et al. 2019. Widespread shifts in the coastal biota of northern California during the 2014–2016 marine heatwaves. *Scientific Reports* 9: 1–14.

Saunders, S. P., N. L. Michel, B. L. Bateman, C. B. Wilsey, et al. 2020. Community science validates climate suitability projections from ecological niche modeling. *Ecological Applications* 30: e02128. DOI: 10.1002/eap.2128.

Schiebelhut, L. M., J. B. Puritz, and M. N. Dawson. 2018. Decimation by sea star wasting disease and rapid genetic change in a keystone species, *Pisaster ochraceus*. *Proceedings of the National Academy of Sciences* 115: 7069–7074.

Schilthuizen, M., and V. Kellerman. 2014. Contemporary climate change and terrestrial invertebrates: evolutionary versus plastic changes. *Evolutionary Applications* 7: 56–67.

Schlegel, F. 1991. *Philosophical Fragments*. P. Firchow, transl. Minneapolis: University of Minnesota Press.

Simpson, C., and W. Kiessling. 2010. Diversity of Life Through Time. In *Encyclopedia of Life Sciences* (ELS). Chichester, UK: John Wiley & Sons. DOI: 10.1002/9780470015902.a0001636.pub2.

Sinervo, B., F. Mendez-de-la-Cruz, D. B. Miles, B. Heulin, et al. 2010. Erosion of lizard diversity by climate change and altered thermal niches. *Science* 328: 894–899.

Spottiswoode, C. N., A. P. Tøttrup, and T. Coppack. 2006. Sexual selection predicts advancement of avian spring migration in response to climate change. *Proceedings of the Royal Society B: Biological Sciences* 273: 3023–3029.

Squarzoni, P. 2014. *Climate Changed: A Personal Journey Through the Science*. New York: Abrams.

Stinson, D. W. 2015. Periodic status review for the brown pelican. Olympia, WA: Washington Department of Fish and Wildlife. 32 + iv pp.

Tape, K. D., D. D. Gustine, R. W. Ruess, L. G. Adams, et al. 2016. Range expansion of moose in Arctic Alaska linked to warming and increased shrub habitat. *PloS One* 11: e0152636.

Telemeco, R. S., M. J. Elphick, and R. Shine. 2009. Nesting lizards (*Bassiana duperreyi*) compensate partly, but not completely, for climate change. *Ecology* 90: 17–22.

Teplitsky, C., and V. Millien. 2014. Climate warming and Bergmann's Rule through time: is there any evidence? *Evolutionary Applications* 7: 156–168.

Teplitsky, C., J. A. Mills, J. S. Alho, J. W. Yarrell, et al. 2008. Bergmann's Rule and climate change revisited: disentangling environmental and genetic responses in a wild bird population. *Proceedings of the National Academy of Sciences* 105: 13492–13496.

Terry, R. C., L. Cheng, and E. A. Hadly. 2011. Predicting small-mammal responses to climatic warming: autecology, geographic range, and the Holocene fossil record. *Global Change Biology* 17: 3019–3034.

Thoreau, H. D. 1906. *The Writings of Henry David Thoreau: Journal, Vol. VIII, November 1, 1855–August 15, 1856.* B. Torrey, ed. Boston: Houghton Mifflin.

Thoreau, H. D. 1966. *Walden and Civil Disobedience.* New York: W. W. Norton & Company.

Tyndall, J. 1861. The Bakerian lecture: on the absorption and radiation of heat by gases and vapours, and on the physical connexion of radiation, absorption, and conduction. *Philosophical Transactions of the Royal Society of London* 151: 1–36.

Vander Wall, S. B., T. Esque, D. Haines, M. Garnett, et al. 2006. Joshua tree (*Yucca brevifolia*) seeds are dispersed by seed-caching rodents. *Ecoscience* 13: 539–543.

Veblen, T. 1912. *The Theory of the Leisure Class.* New York: The Macmillan Company.

von Humboldt, A. 1844. *Central-Asien.* Berlin: Carl J. Klemann.

von Humboldt, A., and A. Bonpland. 1907. *Personal Narrative of the Travels to the Equinoctial Regions of America During the Years 1799–1804,* vol. II. London: George Bell & Sons.

Waldinger, M. 2013. Drought and the French Revolution: The effects of adverse weather conditions on peasant revolts in 1789. London: London School of Economics, 25 pp.

Wallace, A. R. 2009. "On the Law Which Has Regulated the Introduction of New Species (1855)." *Alfred Russel Wallace Classic Writings*: Paper 2. http://digitalcommons.wku.edu/dlps_fac_arw/2.

Weiss, L. C., L. Pötter, A. Steiger, S. Kruppert, et al. 2018. Rising CO_2 in freshwater ecosystems has the potential to negatively affect predator-induced defenses in *Daphnia*. *Current Biology* 28: 327–332.

Welch, C. A., J. Keay, K. C. Kendall, and C. T. Robbins. 1997. Constraints on frugivory by bears. *Ecology* 78: 1105–1119.

White, G. 1947. *The Natural History of Selborne*. London: The Cresset Press.

Wilf, P. 1997. When are leaves good thermometers?: A new case for leaf margin analysis. *Paleobiology* 23: 373–390.

Wilson, W. 1917. *President Wilson's State Papers and Addresses*. New York: George H. Doran Company.

Wisner, G., ed. 2016. *Thoreau's Wildflowers*. New Haven, CT: Yale University Press.

Wodehouse, P. G. 2011. *Very Good, Jeeves!* New York: W. W. Norton & Company.

Woodroffe, R., R. Groom, and J. W. McNutt. 2017. Hot dogs: high ambient temperatures impact reproductive success in a tropical carnivore. *Journal of Animal Ecology* 86: 1329–1338.

Wronski, T., and B. Hausdorf. 2008. Distribution patterns of land snails in Ugandan rain forests support the existence of Pleistocene forest refugia. *Journal of Biogeography* 35: 1759–1768.

Yao, H., M. Dao, T. Imholt, J. Huang, et al. 2010. Protection mechanisms of the iron-plated armor of a deep-sea hydrothermal vent gastropod. *Proceedings of the National Academy of Sciences* 107: 987–992.

Zak, P. 2012. *The Moral Molecule: How Trust Works*. New York: Plume.

Index

acclimatization, 108
African goshawk, 69, 70
AI. *See* artificial intelligence
Aleutian Islands, 56
algae
    bladder wrack, 148
    brown, 148
    coralline, 150
    fossil fuel from, 23–24
    marine (death en masse), 23
    movement of (Northern
        California), 61
    rockweed, 148, 149
    symbiotic, 53, 112
Andrena astragali, 42
aragonite
    addition of layers, 81
    phase vulnerability, 79
    shells formed from, 78
Aristotle, 7, 85
Arrhenius, Svante, 15
artificial intelligence (AI),
    164
attractiveness, sexual selection
    and, 124
Austen, Jane, 174

Beagle, 10
bears
    behavioral shift of (as form of
        plasticity), 108
    berries consumed by, 105
    "contingent interaction" with,
        102
    GPS collars on, 104
    North American brown bear
        (grizzly), 101, 103
    polar bears, 77, 79, 175
    salmon and, 103, 104
Beechey, Frederick, 175
beetle entrepreneur, 65
Bering Sea, 56
biodiversity
    alongside Kodiak's streams, 106
    hotspot, 70
    safe havens for, 143
    shifting, 92
    specialization and, xv
birds
    arrival and disappearance of
        (Europe), 85
    bird-dispersed plant species, 99
    blue jays, 97, 98, 99

birds (*continued*)
    breeding plumage in, 125
    brown owls, 123
    chicks residing in living room,
        25
    Christmas Bird Count
        (Audubon Society), 163, 164
    collared flycatcher, 125, 126
    cormorants, 59, 60
    counting of, 58
    decline of (in smaller patches),
        71
    dovekies, 175, 177, 178
    European nightjar, 39
    fish-eating, 59
    flightless, 192
    flock leaving cliffside nesting
        colony, 175
    foraging of, 38
    of Galápagos, 66, 123
    golden-crowned kinglets, 162,
        163, 164, 167
    great green macaw, 3, 4, 5, 6
    gulls, 59, 60, 173
    hawks, 137
    high elevation specialist
        (Peru), 76, 77
    hummingbirds, 39, 44
    little auk, 175, 176, 178
    long-distance transport by, 98
    migration (White's perspective
        on), 86, 88
    mountain forest (challenges
        facing), 70
    North American (shift of
        winter ranges in response to
        climate change), 100
    pelicans, 58, 59, 65, 174
    people coexisting with, 165
    raptors, 69
    response to temperature
        change, 73, 166
    song sparrows, 150
    songbirds (salmon-derived
        nutrients in), 106
    species living in predicted
        spaces, 168
    swallows, 88
    tawny owls, 123, 124
    Thoreau's observations at
        Walden Pond, 32
    timing mismatches, 39
    turkey vultures, 137
    upslope migration (Papua New
        Guinea), 72, 75
    vagrant (sighting of), 58
    white-crowned sparrows, 71
    woodpeckers, 38, 67
birdsong, as Nature's grandest
    voice (Thoreau), 38
bivalves
    periostracum of, 81
    reliance on aragonite, 78
Black, Joseph, 18
bottlenose dolphin, 61
Bradbury, Ray, 187
Bristol Cliffs, 136, 139–142, 144
British Antarctic Survey, 79
Bunnell, Fred, 63
butterfly effect, 173, 180, 187
Bwana Ndege, 71

calcite, 78
calcium carbonate
    assembly from seawater, 78
    in shellfish, 79

Callophrys eryphon, 68
Canadian Arctic, 80
carbon dioxide
    atmospheric levels during
        Paleocene-Eocene Thermal
        Maximum, 193, 194
    climate change and, 17
    earth's surface temperature
        and, 15
    experiments, 21–22, 26
    fermentation-produced, 19, 21
    molecule, illustration of, 17
    ocean acidification and, 78
    photosynthesis and, 23
    SPRUCE field station
        experiment using, 169–170
    surge of tree growth and, 202
    ubiquity of, 18
carbonated water, discovery of,
    20
carbonic acid, 79
Carver, Thomas Nixon, 101
cascading effects, 106
catastrophism, 10
CEMs. *See* climate envelope
    models
Centrostephanus rodgersii, 90
chaos theory, 173
Cheshire quartzite, 136
Chimborazo diagram (Von
    Humboldt), 156, 158
Christmas Bird Count (National
    Audubon Society), 100,
    163
climate change. *See also* carbon
    dioxide; global warming
    addressing, recommendations
    for, 212

advice for concerned citizens,
    211
animals in motion in era of, 60
Arrhenius's thoughts on, 15
behavior modification and, xvii
biology, xvi, 27, 106, 122, 213
buffer from, 143
carbon dioxide and, 17
challenge in predicting results
    of, 174, 187
coral disease and, 133
dandelion plasticity and, 109
disconnect about, xiv
early warning sign of (lizards),
    46
effects on relationships, 37
effects witnessed on Kodiak
    Island, 103
environment put on different
    trajectory, 147
evolving traits amid, 115
extreme weather swings and,
    44
favoring omnivores and
    generalists, 107
first international treaty on,
    xiii
flexibility of species amid, 37
forest ecosystems impacted by,
    95
forest loss and, 71
as global public concern, xiv
golden-crowned kinglets
    affected by, 164, 167
hallmarks of, 27
heightened aggression and, 204
ice margins vulnerable to
    (Arctic Ocean), 175

climate change (*continued*)
  iconic scenario (polar bears
    and sea ice), 77
  impacts on ocean, 52
  improved germination and, 100
  internal combustion engines
    and, 208
  key fact (plants), 23
  leaves as link to, 192
  lizards and (questions), 47
  modeling alternatives,
    168–169
  most important premise of, 26
  movement in response to, 96
  "Nostradamus of," 48
  ocean acidity and, 78
  pace of, 142
  phenomena interfering
    with natural responses to,
    186–187
  projections, 166
  prolonged responses triggered
    by, 123
  questions about the future and,
    151
  rapid (search for past),
    197–198
  response of trees to, 93, 96, 99
  shift of winter ranges in
    response to (North
    American birds), 100
  shifting of tree species with, 96
  species addition and, 56
  species movement and, 92
  speed of, xv
  as threat multiplier, 205
  transition to global public
    concern, xiv

  trend now accelerating with,
    183
  in the tropics (question about),
    73
climate envelope models
    (CEMs), 161
CNRS. *See* French National
    Centre for Scientific
    Research
cold-blooded animals, 46
Cole, Ken, 181, 183, 185
Columbia River
  pelican populations on, 59, 65
comfort zones
  conditions at edge of, 200
  description of, 45, 47
  habitat variety and, 46
  in rapidly warming word, 201
  starfish, 51
computational ecology, 94
Congo Basin (Africa), 141
Conrad, Joseph, 29
Copley Medal (Priestley), 20
corals
  bleaching of, 112, 113, 114
  death of (ripple effects of), 53
  declines of, 77
  digestion of, 115
  disease and starvation, 113
  movement among reefs, 169
  polyps, 112
  reef-building, 112
  reliance on aragonite, 78
  tropical (decline of), 53
Cretaceous period, 193
critical thermal maximum, 46
culturing (fermentation), 21
Cuvier, Georges, 9, 11

Dacia, 14
Daniel, Alicia, 136, 139, 142
Darwin, Charles, 10, 11, 66, 77, 124
Deacy, Will, 103, 106, 111
decision trees, 165
Dendroctonus, 63
Des Moines River valley (Iowa), 93
Dessler, Andrew, xvi
di Lampedusa, Giuseppe Tomasi, 117
Diamond, Jared, 72, 73, 74
Diatryma, 192
dinoflagellates, 112
disease
 coral, 113
 European (American colonial populations decimated by), 202
 malaria, 74
 marine outbreaks, 52
 prevalence of (heat and), 50, 53
 starfish, 56
 "strange bedfellows" conundrum and, 91
dissertation defense (biology), 74–75
DNA, 83
 analysis tools, 123
 ancient, recovery and analysis of, 200
 cutthroat trout, 132
 Neanderthal, 130
 rainbow trout, 131
Donihue, Colin, 117, 118, 121, 122
Don't Even Think About It, xiv

Earth Summit (1992), xiii, xiv
ecosystems
 climate change events in, 27
 climate impacts on one species affecting, 56
 forest (impact of climate change on), 95
 human-driven trends altering, 186–187
 marine (aragonite and), 78
 novel combinations and communities of, 67, 91
 phenological changes in, 39
 restructuring of, xv
 ripple effects of coral death on, 53
 terrestrial (defining of), 158
 vulnerability of (study of), 200–201
 warming experiment, 169
Eldredge, Niles, 12, 13
environmental determinism, 201
Eocene epoch
 as case study in global warming, 194
 fossils, 192, 195
 marine plankton during, 199
 sandstone, 191
evolution
 biology and, 5
 climate-driven, 128–129
 conditions leading to, 13
 documentation (gold standard for), 123
 fossil record and, 12
 happenstance and, 127
 hurricanes driving, 121
 hybridization, 130–132

evolution (*continued*)
  lizard, 46
  measurement of, 82
  by natural selection (papers
    on), 11
  punctuated equilibrium and, 13
  rapid (competition and), 125
  in response to weather, 122
  tangible signs of (measurement
    of), 115
Evolutionary Applications, 115
extinction
  conditions leading to, 13
  "escalator to," 74, 75–76
  events (Cuvier), 10
  events (major), 194
  as fate of all species, 5
  global temperature variations
    and, 196
  hybridization vs., 132
  local, 77
  lost species and (Cuvier), 9
  major climate shifts and, 195
  mass, 194, 199
  of non-avian dinosaurs, 190
  overhunting and, 14
  of Pleistocene megafauna, 202
  Pleistocene Overkill and, 184
  warning systems for, 200

Fei, Songlin, 96, 99, 100
fermentation, 21, 27
Field Naturalist Program, 135
fish
  anchovies, 59
  butterfly fish, 111, 113, 114,
    115
  hatchery-raised, 130
  hoodwinker sunfish, 61
  kokanee, 133
  marine schooling, 59
  rainbow trout, 129, 130, 131
  salmon, 59, 101, 102, 103,
    104, 133
  sardines, 59
  three-spined stickleback, 126
  westslope cutthroat trout, 128,
    130, 131
Flathead River Valley
  (Montana), 129
Fordham, Damien, 200, 201
Forest Inventory and Analysis
  Program, 95, 96
Fort, Jérôme, 177
fossils
  climate record of, 196
  collections, search of, 200
  Eocene, 192, 193
  evidence, of sloths, 185
  fuels, 24, 107, 210
  gastropods, 81
  hunting, 190, 191
  Joshua tree, 182
  marine, discovery of (Industrial
    Revolution), 9
  plankton communities,
    reconstruction of, 80
  punctuated equilibrium of,
    12–13
  record (communities
    disappeared from), 10
  record (observations from),
    195
  record (of tree advancement),
    99
  species, appearance of, 12

Franz Josef Land, 175, 176, 178, 179

Freeman, Alexandra Class, 72

Freeman, Ben, 71, 74, 75, 77, 92

French National Centre for Scientific Research (CNRS), 176

fruit bats, 5, 50

Fucus, 148, 149

Fucus guiryi, 149

Galápagos Islands
  birds of, 66, 123
  Darwin's objective for visiting, 10–11
  tameness of the birds in, 66

Galileo, 17

GAMs. *See* generalized additive models

gastropods (fossil), 81

generalized additive models (GAMs), 161

generalized linear models (GLMs), 161

genetic drift, 127

Gladwell, Malcolm, 90

GLMs. *See* generalized linear models

Global Crisis, 202–203

Global Marine Hotspots Network, 89

global warming, xiii. *See also* climate change
  case study in, 194
  early Eocene as case study in, 194
  early use of phrase, xiii–xiv
  embrace of life zones, 161
  impacts of, 213
  intense storms and, 122
  rate of, 171
  weather extremes and, 54

"global weirding," 60–61

gorillas, 141

Gotland Island (Sweden), 125

Gould, Stephen Jay, 12, 13

Grand Canyon, 185

Grant, Peter, 123

Grant, Rosemary, 123

"Great Chain of Being," natural world as, 8

greenhouse gases, xvi, 194, 197

Greenland, temperatures over, 198

Grémillet, David, 176, 179

habitat
  beetles colonizing, 65
  chemically unlivable, 82
  climate shifts and, 199
  Costa Rica, 160, 165
  disappearing (birds), 74
  dwindling, 127
  elevation and, 156, 157
  elfin forest (Peru), 76
  expansion of concept of, 77
  fringe populations and, 200
  introgression and, 130–131
  Joshua tree, 181
  jungle, 155
  lake (shrinking), 133
  loss (ecosystems altered by), 186–187
  loss (kinglet), 167
  in miniature, 139

habitat (*continued*)
  model (golden-crowned
    kinglets), 164
  mountain (American pika),
    143–145
  nesting (death camas bees), 42
  new arrivals to, 59
  Ouija board, 158, 159
  shifting uphill, 144
  SPRUCE project, 170
  unsuitable (for trees), 67
  variety (comfort zones and), 46
Hale, Edward Everett, 211
Harvell, Drew, 50, 51, 52, 56
heliotherms, 49
Hell's Half Acre (cliffs), 136
Holdridge, Leslie, 154, 157,
  160
Holdridge Life Zone System, 154,
  159
Homo sapiens, 46, 132
hours of restriction, 49
Humboldt squid, 110, 111
Hurricane Irma, 118
Hurricane Maria, 118
Hutton, James, 7, 9, 10
hybridization
  climate-driven, 131
  description of, 130
  extinction vs., 132
hydrology, 77

Industrial Revolution, 9
insects
  death camas bees, 41–44
  emergence of, 35
  expected hatch, 39
  ground-nesting bees, 43

mountain pine beetle, 62, 63,
  65, 66
red-eyed fruit fly, 17
speckled wood butterflies, 123
western pine elfin, 68
internal combustion engines, 208
intertidal zone, 53
*Introduction to Modern Climate
  Change*, 16
introgression, 130, 131

Jefferson, Thomas, 9
jellyfish, 112
Joshua tree, xii, xiii
  ancient partnership of sloths
    and, 185, 186
  Cole's project on, 181
  fossils, 182
  fruits, 184, 185
  history of, 182
  response to temperature, 183
  seed disperser, 186, 199
  unexpected movement of, 186
Joshua Tree National Park, xii,
  181
Justice League, 107

Keith, Sally, 111
keystone species, 53, 56
King John, 207
King Lear, xi
Kodiak Island (Alaska), 103,
  104, 106
Kolka, Randy, 169, 170
Kovach, Ryan, 132

La Selva Biological Station
  (Costa Rica), 153–156, 161

Laurenço, Carla, 147, 148
leaf margin analysis, 192, 193
Leopard, The, 117
Leopold, Aldo, 39
Lewis and Clark Expedition
    (1804), 9
life zones, 155
    basis of, 156
    of Costa Rica, 160
    global warming studies'
        embrace of, 161
    Holdridge Life Zone System,
        154, 156, 159
    introduction of concept of, 160
    most prominent, 158
    Von Humboldt's basic insight
        and, 157
Limacina helicina, 80
Lindgren, Staffan, 63, 64, 65,
    66, 92
Lindgren funnel trap, 65
Linnaeus, Carl, 7, 86
Little Ice Age, 202, 203
lizards
    anole family, 117, 120, 122
    biological responses of, 212
    capture, 49, 119
    climate change and
        (questions), 47
    effects of hurricanes on
        populations, 118, 121
    evolution, 46
    fence lizards, 48
    hours of restriction, 49
    hurricane simulation with, 119
    research on, 117
    spiny lizards, 48
long-spined sea urchins, 90

Lord of the Rings, The, 93
Lorenz, Edward, 173
lost species, 9
Love, Renee, 190, 192, 195, 199
Lyceum, 189
Lyell, Charles, 10

machine learning, 164
Mackay, Charles, 85
MAD. *See* Move, Adapt, or Die
Marshall, George, xiv
Marsham, Robert, 39
Martin, Paul S., 184
Martin, Thomas-Henri, 17
mass extinction, 194, 199
mate choice, 124. *See also* sexual
    selection
McCrea, William, 173
megafauna, 184, 202
mephitic air, 20. *See also* carbon
    dioxide
methane, 17, 194
microbial digestion, 21
microrefugia, 143
midden, 182, 183
Migrationes Avium, 86
mining bees, 42
Mount Karimui (New Guinea),
    72–74
Move, Adapt, or Die (MAD), 83
movement (of species)
    adaptive movement (Peel), 89
    of algae (Northern California),
        61
    at Bristol Cliffs talus, 142
    climate change and, 92, 96
    of corals among reefs, 169
    Joshua trees, 186

movement (of species)
(*continued*)
    meaning and details of, 91
    measurement by tracking
        geographic center, 96
    ocean warming and, 61
    reasons for, 92
    shifting conifers replaced by
        hardwoods, 139
    simultaneous adaption and,
        100
    of trees, 97, 99
    White's inquiries into, 86
mutualism, nature's best-known
    example of, 112

Natural History of Selborne, The,
    85
nature, driving forces of (White),
    88
Newmark, Bill, 70, 71
Newton, Isaac, 93
nonlinear relationships, 180

oceans
    acidification, 78, 79, 82
    Arctic Ocean, 175, 177
    Atlantic Ocean, 198
    cool currents in, 140
    currents, cyclic changes in, 202
    distribution of cephalopods
        in, 89
    flow of energy from, 106
    glacial melt and, 178
    impacts of climate change on,
        52
    Pacific Ocean, 144
    sediments, 9

upper-latitude, 80
    warming (algae-rich waters
        of), 126
    warming (disease and), 53, 55
    warming (species movement
        and), 61
    warming (stressed organisms
        and), 52
Ochotona princeps, 145
Oncorhynchus clarkii lewisi, 131
orcas, 57, 58
Orians, Gordon, 211
Origin of Species, The, 11, 12
osmotic shock, 177
oysters
    commercial farms, 79
    larval, 78
    vulnerability to acidity, 82

Pack Creek Bear Viewing Area
        (Alaska), 101
Paine, Robert, 53
Paleocene-Eocene Thermal
        Maximum, 194, 195
parameter-elevation relationships
        on independent slopes
        models (PRISMs), 161
parasites, 66
Parker, Geoffrey, 202
Parmenides, 7
pathogen, 51, 52, 66
Peale, Charles Willson, 11
Peck, Victoria, 79, 82
Pecl, Gretta, 89, 92, 100, 167
periostracum, 81
photosynthesis, 18
    carbon dioxide and, 23
    of plankton, 112

Pisaster, 53, 54, 56
Pisaster ochraceus, 55
plankton
  abrupt transition that killed,
    177
  bottom-dwelling marine, 195
  communities (fossilized), 80
  coral polyps feeding on, 112
  curtains, 177, 179
  photosynthesis of, 112
  pteropods feeding on, 80
  reliance on aragonite, 78
plants
  alpine (experimental shifting
    to lower elevations), 169
  Arctic tundra, 67
  berries, 104, 105–106, 172
  cacao, 154
  dandelions, 109, 110
  death camas, 41, 42, 43
  ferns, 159, 190, 192
  flowering (temperature as
    driver of), 35
  fruit, 39, 93
  grasses, 41, 67, 103
  Labrador tea, 137
  leaf margin analysis, 192
  leatherleaf, 137
  leaves as link to climate
    change, 192
  lichen, 44, 137
  paramo (grass zone), 160
  peach palm, 154
  violets, 35
  wood sorrel, 35
Plastic Man, 107
plasticity
  body size and, 111

  climate stability and, 114
  of dandelions, 110
  deficiency of, 115
  description of, 107
  of dietary changes, 179
  distribution of, 109
  importance of, 212
  permanent alterations
    produced by, 108
  unexpected, 168
Plato, xiv
Pleistocene epoch, 141, 146, 182
Pleistocene Overkill, 184, 199
Plexiglas windows, 107
Pliny the Elder, 85
pollen records, 99
polyp, 53, 112
pre-montane transition
    (rainforest), 159
Priestley, Joseph, 18, 19, 24, 78
Primack, Richard, 32, 33, 40, 44
Principles of Knowledge, 9
PRISMs. *See* parameter-
    elevation relationships on
    independent slopes models
pteropods, 80
punctuated equilibrium, 12, 13
Pycnopodia, 51, 55, 56

QED moment, 213
Quality Comics, 107
quartzite, 136

rainforest
  canopy, 50
  Congo Basin, 141
  extinction escalator and,
    75–76

rainforest (*continued*)
  at La Selva, 155
  lowland, 4, 155
  pre-montane transition and,
    159
Random Forest, 165
Range Extension Database and
    Mapping Project, 89
raptors, 69
refugium, 141, 143
Reid, Clement, 97, 98
Reid's Paradox, 98
Roby, Dan, 58, 59, 60, 61,
    65, 92

Sagan, Carl, xv
San Bernardino Mountains, 109
Sceloporus, 48
Schlegel, Friedrich, 189
Schweppe, Johann, 20
sea stars, 55. *See also* starfish
  colorful, 54
  ochre, 55
  predatory, 53
  sunflower, 51, 56
sexual selection
  attractiveness and, 124
  definitive change in, 127
  economics of, 126
  of songbirds, 125
Shakespeare, William, xi, 57, 207
Sharpe's akalat, 69, 70
Shasta giant ground sloths, 184,
    185, 186, 199
shellfish, calcium carbonate in,
    79
Sierra Nevada mountains
    (California), 144

Sinervo, Barry, 46, 47, 48, 50
snails
  periostracum of, 81
  reliance on aragonite, 78
  sea butterflies, 79–82
Socrates, xv
spare simplicity (Thoreau), 33
Species on the Move, 89
species movement
  adaptive movement (Peel), 89
  of algae (Northern California),
    61
  at Bristol Cliffs talus, 142
  climate change and, 92, 96
  of corals among reefs, 169
  Joshua trees, 186
  meaning and details of, 91
  measurement by tracking
    geographic center, 96
  ocean warming and, 61
  reasons for, 92
  shifting conifers replaced by
    hardwoods, 139
  simultaneous adaption and,
    100
  of trees, 97, 99
  White's inquiries into, 86
Sphagnum moss, 137
spiders, 106, 122
SPRUCE (Spruce and Peatland
    Responses Under Changing
    Environments), 169–171,
    180
starfish. *See also* sea stars
  comfort zone, 51
  death of (ripple effects of), 53
  disease, 56
  purple, 54

species dying, 51
tank full of (accidental heating of), 50, 51
"survival of the fittest," 120, 124
symbiont, 112

talus, 136, 137, 138, 140, 147
tapir, 192
Taraxacum officinale, 110
Teflon, 107
Tempest, The, 57
Tertiary Period, 193
Theory of the Earth, 7
Theory of the Leisure Class, The, 3
thermostat manufacturers, 47
Thoreau, Henry David, 31, 32, 33
Three Men in a Boat, 153
timing mismatches, 39, 43, 44
Tolkien, J. R. R., 93
Toxicoscordion venenosum var. *venenosum*, 41, 42
Trajan (Roman emperor), 14
trees. *See also* Joshua tree
    advancement, fossil and pollen records of, 99
    alders, 191
    almendro, 4, 5, 6, 159
    beetle damage to, 62
    birches, 191
    bog (SPRUCE project), 180
    breaking bud early, 172
    conifers, 136
    crops (experiments on), 154
    damaged (rapid regrowth of), 122
    deciduous, 37
    Douglas fir, 62
    pine elfins and, 68

environmental conditions for sprouting and growing, 94
    flexible, 37
    Forest Inventory and Analysis Program, 95
    healthy (infestation of), 63
    hickories, 94, 137
    immobile, 93
    lodgepole pines, 62
    long-distance movement in (dynamics of), 97
    maples, 94, 136, 137, 141
    mass reforestation of abandoned farmland (colonization of the Americas), 202
    measurement of populations, 96
    as naïve hosts, 65, 66
    oaks, 32, 97, 99, 141
    panting, 5
    pine (regeneration of), 67
    pine (sapwood in), 61
    pine (skeletal forests of), 64
    relocation (bird relocation vs.), 100
    response to climate change, 93
    roost (fruit bats), 50
    species, shifting of, 96, 97
    survival (temperature and), 156
    unexpectedly rapid dispersal (examples), 98
    Yggdrasil, 93
Triceratops, 190
Tyndall, John, 24, 25, 26
Typhoon, 29

upslope migration
  birds of Papua New Guinea,
    72
  documentation of, 76
US Forest Service, 62, 95, 101,
    144
US Geological Survey, 59
US National Park Service, 181
Usambara Mountains (Tanzania),
    70, 71, 77

Veblen, Thorstein, 3
Volcan Barva (Costa Rica), 160
Von Humboldt, Alexander, 14,
    156, 157, 158

Walden, 31, 32
Walden Pond, 31
  average temperatures around,
    35
  bean field planted at, 37
  blocks of ice exported from, 36
  changes at, 32

Primack's focus on, 34
Thoreau's observations at, 32,
    33
Wallace, Alfred Russel, 11
warm-blooded animals, 46
whale-watching boats, 57
White, Gilbert, 85, 86, 88
Wilsey, Chad, 164, 165, 168
Wilson, Woodrow, 1
Wodehouse, P. G., 69
Wright, Frank Lloyd, 154

Yggdrasil (tree), 93
yucca. See also Joshua tree
  fruits unusual for, 184
  world's largest variety of, xiii
Yukon Territory (Canada), 66

zooxanthellae, 112
zooplankton
  charisma of, 79–80
  description of, 175
zygacine, 41

Thor Hanson is a biologist, Guggenheim Fellow, and author of award-winning books including *Buzz*, *Feathers*, *The Impenetrable Forest*, and *The Triumph of Seeds*. He lives with his wife and son on an island in Washington State.

www.thorhanson.net